Design of High Pressure Steam and High Temperature Water Plants

DESIGN OF HIGH PRESSURE STEAM AND HIGH TEMPERATURE WATER PLANTS

WILLIAM L. SCULTHORPE, Project Manager
Jaros, Baum & Bolles, Consulting Engineers

INDUSTRIAL PRESS INC., 200 Madison Avenue, New York, N. Y. 10016

Contents

Preface

The young mechanical designer of today is well based in theory but usually lacking in the area of practical design of high pressure steam and high temperature water plants. He may acquire practical design information in the drafting room, and from technical books which contain a tremendous amount of information, but generally not on a step-by-step basis leading to a final design. This book was written on a practical "how-to-design" basis for the designer, and it is the author's hope that this small contribution will be of help.

To his good friend, Richard L. Koral, editor and publisher of BUILDING SYSTEMS DESIGN, the author wishes to express his warmest thanks for his encouragement and great help with the manuscript, and the publishing of material from this book in BUILDING SYSTEMS DESIGN. The author is indebted to Georg Stabenow, Russel Imbt, Jr., E. D. Frick and Milton L. Greenberg of the International Boiler Works Co., a subsidiary of Ovitron Corp., for their great assistance in preparing the section on high temperature water. For their very valuable suggestions and assistance the author wishes to express his thanks to the following people: Irwin S. Feldman, Yula Corporation; William C. Freed, Cochrane Products, Crane Co.; Robert G. Eberly, Detroit Stoker Co.; J. A. Wagner, Electronics Corp. of America; Lorenzo Burrows, Cleaver-Brooks Div., Aqua-Chem, Inc.; H. S. Johnson, Buffalo Forge Co.; A. H. Seidler, Todd-CEA; Chris J. Vorndran, Combustion Equipment Associates, Inc.; Salvatore F. Lieberman, Bailey Meter Co.; Richard T. Blake, Metropolitan Refining Co., Inc.; and J. J. Coffey, Murray Iron Works Co.

The author also wishes to express his thanks to the following for their generous assistance in supplying information, and/or illustrations: The Babcock & Wilcox Company, for much of the material on steam generators, pulverizers and chimneys that has been taken from their book STEAM—ITS GENERATION AND USE; National Coal Association and Gifford-Wood Inc., for coal and ash handling; Todd-CEA, for fuel burning equipment and systems; Yula Corporation, for heat exchangers; Metropolitan Refining Co., Inc. and Cochrane Products, Crane Co., for water treatment; Bailey Meter Co., for combustion controls and instrumentation; Electronics Corp. of America, for safety controls for steam and high temperature generators; Detroit Stoker Co., for stokers; Ric-Wil Inc., for underground steel insulated conduit; Tube Turns, Division of Chemetron Corp. and POWER magazine, for charts; American Society of Heating, Refrigerating and Air-Conditioning Engineers, Inc., for expansion loops for steam and hot water piping; and Erie City Energy Div. of Zurn Industries, Inc., Combustion Engineering, Inc., Orr and Sembower, Inc., Superior Combustion Industries, Inc., American Gilsonite Co., Yarway Corporation, and Peabody Engineering Corp., for photographs and sketches.

1

Steam Generating Plants

In this chapter, the intent is not to enter the highly specialized utility and marine fields of steam generation. Instead, there is an attempt to provide the designer with the information necessary to select a steam generator that will fulfill the requirements of load, pressure, fuels, space and costs.

The designer of today's high pressure steam plant is faced with many questions as he starts out on a new project. Some of these are: at what pressure should the plant operate? What size should the steam generators be? How many should be used? Should they be fire tube or water tube, packaged or field erected? What should the design pressure be? Which unit is the most economical? What is the most economical fuel for the locality? Can we use stub stacks, or is there a problem of air pollution necessitating a tall stack or a higher priced, cleaner fuel? Should we burn gas, oil or coal? If we burn coal, should we use an underfeed stoker, a spreader stoker or a pulverizer? How are fly-ash separation, fuel storage and handling, ash handling, removal, storage and disposal going to be accomplished? What shall the water treatment be?

Types of Steam Loads and Operating Pressures

Steam loads and operating steam pressures for different types of structures vary, making it necessary for the designer to have a good working knowledge of them in order to select proper equipment. The following are considerations in selecting equipment for several types of structures.

Hospitals—Various steam loads for a hospital include heating, ventilating, air conditioning, sterilizing, domestic hot water, kitchen and laundry.

If the steam plant is relatively close to the hospital complex, and pressure drop in the steam distribution piping is minimal, it need not be considered. However, if the plant is remote, and there is much pressure drop in the distribution piping, it must be considered.

Steam loads that greatly influence steam plant operating pressure are the laundry ironer and use of steam turbines for driving air conditioning refrigeration compressors. To be efficient and operate at top speed, the ironer requires dry, saturated steam at 125 psig. If the air conditioning refrigeration equipment operates

at low pressure (such as 12 psig) the steam plant operating pressure may be set at 140 psig. However, if steam turbines are employed to drive refrigeration equipment, the operating pressure for efficient turbine performance would have to be set at 230 psig.

Office buildings — Large office buildings (from 40 to 60 stories high) in cities not having access to high pressure district steam, necessarily furnish their own steam supply. These structures are usually air conditioned and their refrigeration compressors must be driven either by an electric motor or steam turbine. For efficient operation the turbine requires steam at 230 psig. If absorption type water chillers are used, steam pressure would be 12 psig.

Hotels—Hotels have steam loads similar to those of hospitals with the exception of a sterilization load. As this is not a controlling factor, steam plant operating pressure will be either 140 psig or 230 psig.

Industrial buildings — Industrial buildings have many of the previously mentioned steam loads and, in addition, a process steam load. Steam pressure required by

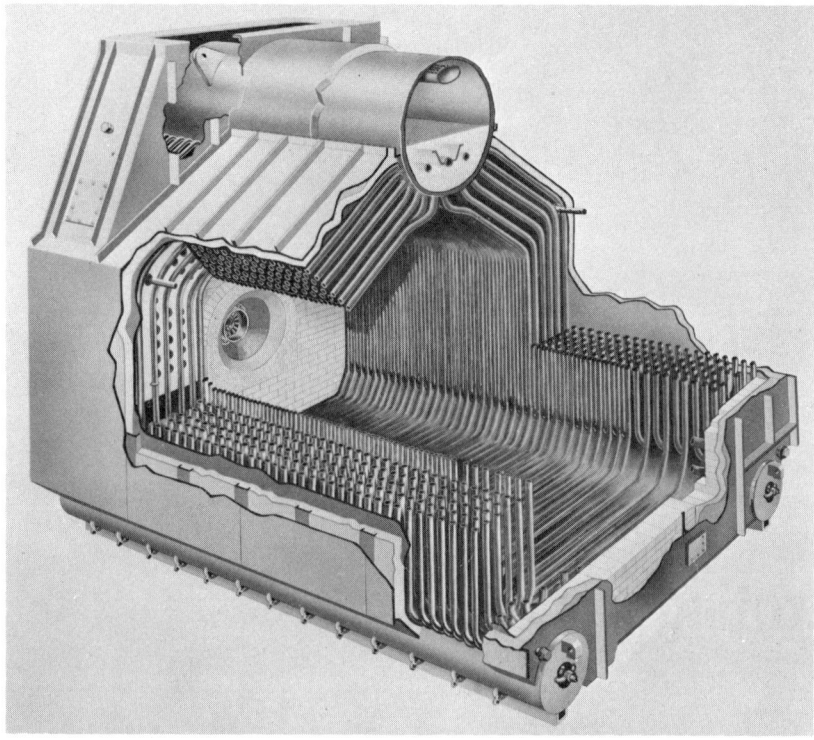

A typical packaged water tube steam generator of the type A design. Courtesy of Wickes Boiler Co.

many types of process steam loads varies greatly and cannot be pinpointed. In determining plant operating pressure, the designer necessarily has to analyze all loads involved and set the pressure accordingly.

Campus-type buildings—The term, campus-type arrangement of buildings, refers to colleges, office buildings, industrial parks, shopping centers, research centers and airports. When a steam plant is used in campus-type construction, it is usually remote from other structures; distribution lines become very long, with attendant large pressure drops. This important factor must not be ignored. After analysis of required steam pressures has been completed, distribution piping pressure drops must be computed in order to determine proper plant operating steam pressure.

Packaged Steam Generators

The packaged steam generator was conceived in the 1930's. However, demand during World War II for packaged, portable, fire and water tube, steam generators

aboard ships, trailers and railroad cars contributed tremendously to the development of these designs. They were considered expendable and were, therefore, fired to their absolute limit in order to produce the greatest amount of steam from the smallest possible unit.

After the war, these units were offered to industrial stationary plants, where long life is considered a prime factor. It was soon recognized that firing must be controlled within certain limits for a reasonably long life. From that time to the present, there have been rapid advances in design of combustion controls, instrumentation, welding techniques, lubrication, assembly procedures, fuel burners, safety controls and feed water treatment.

These compact, high performance, fully automated packages cost less than field erected units, which is one of the reasons for their popularity.

The Packaged, Water Tube Steam Generator

The packaged, multiple drum, natural circulation, water tube

steam generator is available in four designs, namely: types *A, D,* what is generally referred to in the industry as *O,* and the fire box type.

These steam generators are fully automatic, integrated, factory assembled, self-contained units, built upon a structural steel skid-type supporting frame. They are designed to be transported from factory to job by railroad and trailer. Their widths are limited to that of a railroad flat car, and their heights controlled by over-head clearances. Their furnaces are long, narrow and fairly high, which conforms to the design restrictions.

Operating pressures for these units range to 1050 psig, capacities to 200,000 lb per hour. Tube diameters are generally about 2 in. and tube walls, being not too thick, provide a high rate of heat transfer. The radiant water walls produce the majority of the steam, with the convection section completing the job. Fuel burners, either oil, gas or both, have flames shaped to prevent impingement upon the furnace tubes and to fit the furnace length.

The units are positively pressure fired and have either a 10-gage pressure tight steel casing behind the tubes, or a layer of plastic refractory to prevent penetration of sulphur from the fuel oil, through insulation to the 10-gage steel outer casing. Insulation is applied against the water-cooled surface and is generally of a light, not bulky, material.

Safety controls for these steam generators should always comply with requirements of Factory Insurance Association. Combustion controls should be of the full metering type, having proportion and reset, and either electrically or air operated. Instrumentation should include steam flow, fuel gas and fuel oil flow meters, draft gages, flue gas temperature and O_2 and CO_2 recorders, and steam, oil, and fuel gas pressure gages.

The control panel, containing the combustion controls, safety controls and instrumentation, should never be mounted on the unit, but should be set up as a free-standing panel completely clear of the generator. This is to prevent vibration, which is detrimental to this equipment.

If super-heat is required, a super-heater can be incorporated into the generator design.

Where refractory material is required, it should be of the super-duty type.

The feed water regulator should be of the two element type unless the unit has to cope with extra heavy steam load swings, in which case it should consist of three elements.

Water column and gage glass should be of a type to suit the operating pressure.

Blow-off valves should be of the seatless type.

Steam Generator Heating Surface

The heating surface of a water tube steam generator consists of two types—convection and radiant. The radiant surface is that exposed directly to the fire in the furnace. The convection surface is in the steam generator proper, and is exposed only to the hot gases of combustion. All surface is measured on the outside of the tube. The arrangement is as follows:

Tangent—One half the circumference of the tube is exposed to the fire.

Staggered — This arrangement also exposes one half the circumference of the tube to the fire.

Finned—This arrangement presents one half of the circumference of the tube to the fire, plus the finned, or extended, surface.

Spaced Tubes—The full circumference of the tube is exposed to the fire.

In the convector section of the steam generator, tubes are spaced apart to allow hot combustion gases to pass over 100% of their surfaces.

The finned, or extended surface, tube is used only in the radiant section of the steam generator, where a fin, or extended surface machine welded to the tube wall helps to raise tube temperature. Practical limits for this surface are ¼-in. thick by 1-in. wide. Manufacturers rate this surface as 100% heating surface, an excellent supplemental method of heat transfer.

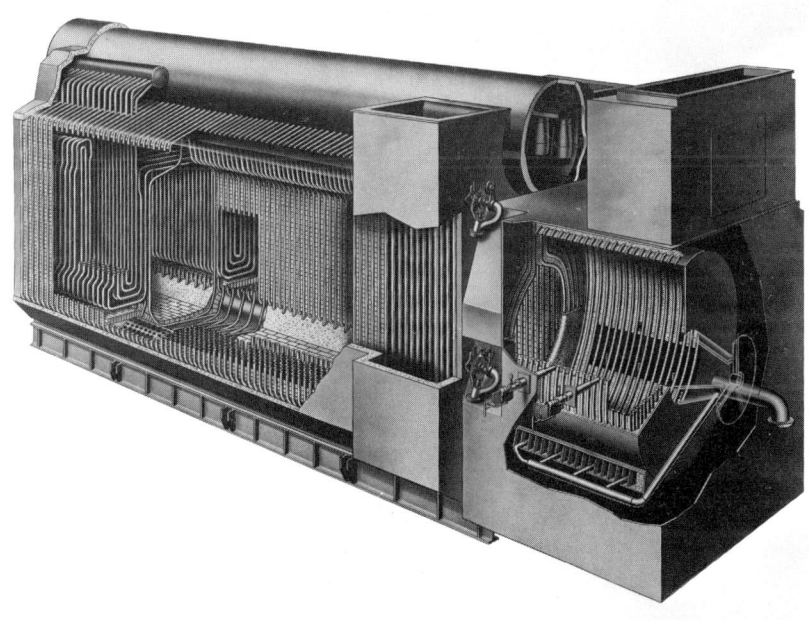

A type A steam generator with a cyclone furnace. Courtesy of The Babcock & Wilcox Co.

The most popular of packaged water tube steam generators is the type D design above. Courtesy of The Babcock & Wilcox Co.

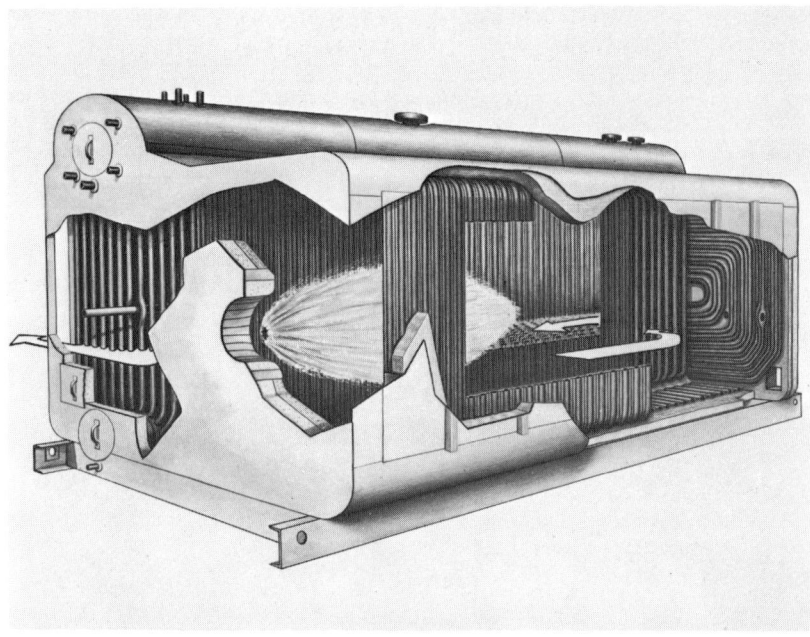

The type D steam generator above has a positively pressurized furnace and is rectangular in shape. Courtesy of Murray Iron Works Co.

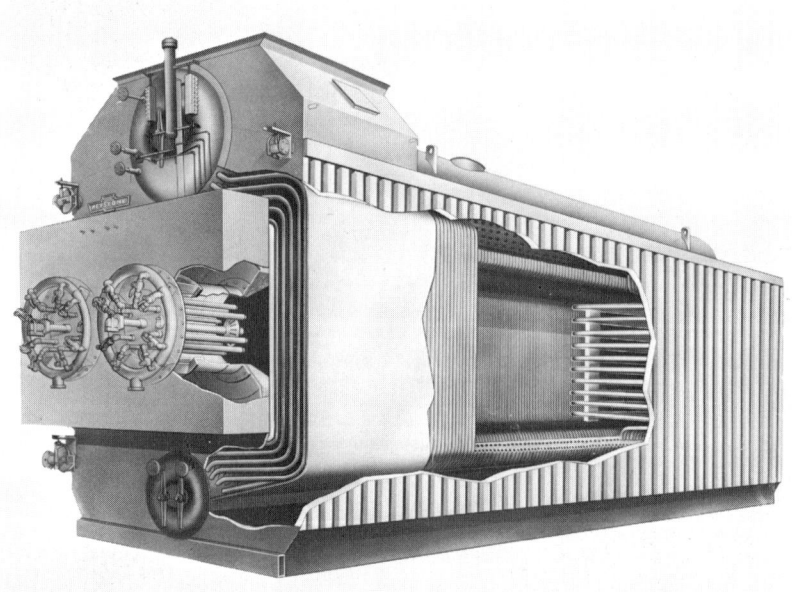

The type O or Keystone steam generator. A removable superheater module can be readily installed or removed through rear of unit. Courtesy of Erie City Iron Works.

Heat Transfer and Release Rates

Average heat transfer and heat release rates in the packaged, water tube steam generator have been a controversial subject for many years. This type of steam generator can only increase in length, and then only to a practical limit, which means that the average heat transfer and heat release rates must rise with increased generator capacity.

Manufacturers' experience over the years proves that with a well designed steam generator, these rates are practical and workable.

Table 1 indicates the rates from 0-200,000 lb per hour.

Generator Selection

A typical example will show how to go about selecting a steam generator. A steam generator with a capacity of 40,000 lb of dry saturated steam per hour is required. Operating pressure is 225 psig; feed water is supplied at 228F; fuel to be used is No. 6 fuel oil, having a heating value of 148,000 Btu per gal, or 18,500 Btu per lb; average heat transfer rate per sq ft of heating surface is 9200 Btuh.

Heating surface for the steam generator may be calculated from the formula:

$$HS = \frac{W_s\,(H - h)}{9200}$$

where

HS = total effective heating surface in furnace and convection section, sq ft;

W_s = steam load, lb per hr;

H = total heat in steam at given pressure and temperature, Btu per lb;

h = total heat in feed water, Btu per lb.

$$HS = \frac{40,000 \times (1200.6 - 196.16)}{9200}$$
$$= 4360 \text{ sq ft}$$

A steam generator having at least 4360 sq ft of effective heating surface should be selected. A check of a typical manufacturer's catalog for a 40,000 lb per hr steam generator indicates the following, all of which is acceptable: Radiant heating surface (from catalog): 900 sq ft; Total heating surface: 4360 sq ft; Furnace volume (from catalog): 1119 cu ft; Boiler output: 40,000,000 Btuh; Boiler input, 40,000,000/80% efficiency: 50,000,000 Btuh; Heat release per cu ft of furnace volume, 50,000,000/1119: 45,000 Btu.

At maximum steam generator output, heat input to the furnace should not exceed 54,000 Btu per cu ft of furnace volume per hour and 100,000 Btu per sq ft of effective radiant heating surfaces per hour. Combustion gas temperature at the furnace exit should not exceed 2100F for an oil-fired unit and 2150F for a gas-fired one.

Type A Steam Generator

The Type *A* generator is designed with a long steam drum centrally located at the top of the unit. Two tube banks slant downward to headers on each side, forming a triangular-shaped furnace. Each tube bank is composed of risers and down comers and, in a sense, form two separate steam generators with a common steam drum. Tubes entering the lower surface of the steam drum protect the drum from furnace heat.

The furnace floor in this design is either water cooled or bare refractory. If water cooled, the tubes are widely spaced, requiring refractory and insulating materials under them. A small disadvantage of this design is the need for periodic replacement of these refractory materials and insulation. Some designs have air cooled roofs.

The insulation and metal skins are typical of the packaged water tube designs. The positive pressure firing arrangement includes a forced draft fan and burners for oil, gas or both. This design is presently built to a maximum capacity of 200,000 lb steam per hour and a working pressure of 1050 psig. One exception to these capacities is a spreader, stoker-fired, packaged, *A* type unit with a capacity to 60,000 lb per hour, made by Wickes Boiler Co.

Type D Steam Generator

The type *D* design is the most popular, the simplest and most economical to construct of all packaged water tube designs. The radiant water wall tubes are tangent and arranged in a straight line, slightly staggered or with extended surface, the latter being very popular with designers. The furnace is rectangular in shape, with the convection section forming either a right or left hand wall.

The large, deep, positively pressurized furnace is ideally suited for single or multiple burners. Like all packaged water tube designs, it must be increased in length to provide more steam capacity. Fuels used are oil, gas or both. Present capacity limits are 200,000 lb steam per hour, with

Table 1. Heat Transfer and Heat Release Rates for Packaged Water Tube Steam Generators

Generator capacity, lb per hour	Average heat transfer rate (based on total heating surface), Btu per sq ft	Heat release in combustion chamber, Btu per cu ft
0- 40,000	9,200	54,000
40,000-100,000	14,000	76,800
100,000-125,000	18,200	143,000
125,000-200,000	20,000	167,000

steam working pressures up to 1075 psig.

Type O Steam Generator

The type *O* design provides a centrally located furnace running the length of the unit. The steam drum sits directly over the mud drum or bottom drum on the center line of the generator. All walls, floor and roof are completely water cooled. Convection tubes are closely spaced. Furnace tube walls and outer tube walls have tangent tubes. This design provides a well shaped furnace for the burning of oil or gas and presents almost 100% black surface to the fire.

The unit, as the other packaged designs, must be increased in length to provide additional capacity. Water circulation is natural. The unit operates under a positive furnace pressure. This steam generator is presently built to a maximum capacity of 250,000 lb steam per hour and a working pressure of 1000 psig. The type is manufactured only by Erie City Iron Works and is known as *Keystone* Steam Generator.

Firebox Type Water Tube Steam Generator

The firebox type, water tube, steam generator is a packaged design that combines a water-walled firebox with a straight-tube box header, but with the steam drum as an integral part of the unit, not separate as in the box header design. Tube access for cleaning or

A 4-pass, high pressure, modified, scotch marine steam generator. Courtesy of Cleaver-Brooks.

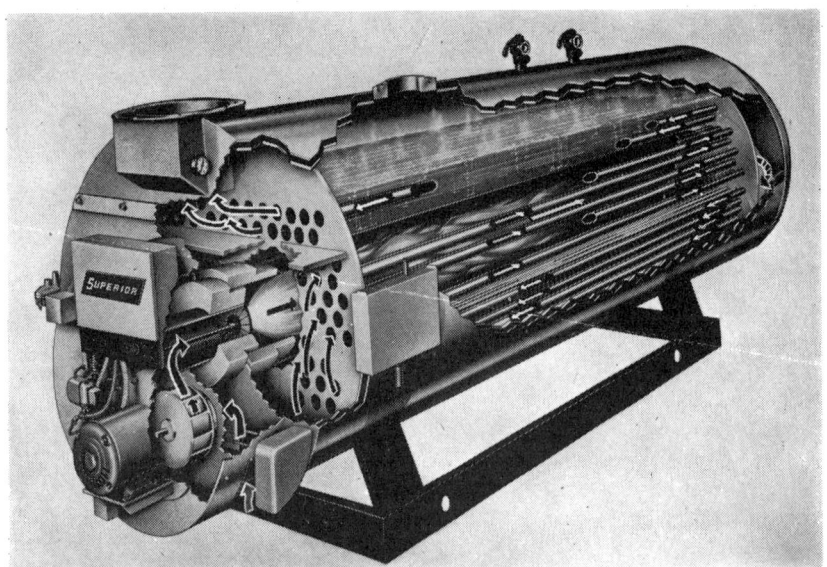

Shipped ready for connection to electric power, steam, feedwater and fuel supplies, is this 4-pass, high pressure, modified, scotch marine steam generator. Courtesy of Superior Combustion Industries Inc.

replacement is through screw plugged openings in the outer front and rear header walls.

This unit has a refractory floor; is designed to burn oil, gas, or both; and can burn coal with an underfeed stoker. A gas or oil-fired unit has a maximum capacity of 26,000 lb steam per hour. A coal-fired unit has a maximum capacity of 10,350 lb steam per hour.

Maximum working pressure of the unit is 200 psig.

Packaged High Pressure Modified Scotch Marine Steam Generator

Packaged, high pressure, modified, scotch, marine, steam generators are designed with dry and wet-back rear walls and arranged for three gas flow patterns, namely: 2-pass, 3-pass and 4-pass. The

3- and 4-pass designs are preferable to the 2-pass as they are easily pressurized, whereas the 2-pass unit requires tube baffling to provide adequate pressure drop. This is not desirable when burning fuel oil, as the tubes easily become blocked with soot and cleaning becomes a problem. The 3- and 4-pass units are so superior that the 2-pass design will probably become nonexistant in the near future.

The shell is cylindrical and houses a large, submerged, internally-fired furnace, with fire tubes arranged around it, providing long, high-velocity gas travel. The generator is arranged and shipped ready for connection to electric power, steam, feed-water and fuel supplies. These units are designed to burn only fuel oil, natural gas and manufactured gas.

Practical shell diameters are limited to approximately 96-in., with a top steam working pressure of 300 psig and maximum tube lengths of 20 ft. Larger shell diameters would require thicker shells at higher cost, which would raise the price of these units above that of water tube steam generators of the same capacity.

Maximum steam capacity for these designs is approximately 24,000 lb steam per hour. Heating surface is set at 5 sq ft per boiler

High Pressure Modified Scotch Marine Steam Generator Selection Chart

Boiler hp 5 sq ft fireside heating surface per bhp 34.5 lb steam from and at 212F	Steam output at nozzle, lb per hr	Fireside heating surface, sq ft	Combustion chamber volume, cu ft (water cooled chamber only)	Heat released, Btu per cu ft in (water cooled combustion chamber only)	Inside drum diameter, in.	Steam space above water line, cu ft (minimum)	Steam disengaging area at waterline, sq ft	Distance from water line to top of steam drum, in.
15	518	75	3.5	180,000	36	2.8	9.2	6½
30	1,035	150	6.4	194,000	36	5	11.7	7
50	1,725	250	13.5	153,500	48	9	20.04	8¾
60	2,070	300	13.5	180,000	48	9	20.04	8¾
70	2,415	350	20	144,000	48	13.5	24.3	8¾
80	2,760	400	20	166,000	48	13.5	27.6	8¾
100	3,450	500	23	180,000	48	15.6	34.4	8¾
125	4,313	625	27.4	189,000	60	21	34.7	10¾
150	5,175	750	33.2	187,000	60	25	43.4	10¾
200	6,900	1,000	40.4	205,000	60	29.8	45.0	13
250	8,625	1,250	63.5	162,000	78	52	54.0	16¾
300	10,350	1,500	77.0	160,000	78	63	66.3	16¾
350	12,075	1,750	92.0	157,000	78	71.9	73.9	16¾
400	13,800	2,000	123.3	138,000	96	82	68.9	20¼
500	17,250	2,500	154.6	135,500	96	103.3	86.8	20¼
600	20,700	3,000	187.8	132,000	96	124.5	104.7	20¼
700	24,150	3,500	208.5	127,500	96	135.92	124.6	20¼

hp, producing 34.5 lb steam per hour. Heating surface is calculated on the fire side only.

Furnaces are positively pressure fired, which accelerates burning of the fuels. Heat release rates, compared to those of packaged water tube steam generators, are very high, but so are rates of heat absorption into the water without a dangerous amount of turbulence in the generator. Large steam disengaging surface and adequate steam space above the water lines, coupled with low steam leaving velocities, produce a tolerable steam quality that, at maximum, is 0.75% moisture in the total amount generated.

The distance from water line to inside top of the steam drum allows only for insertion of a dry plate under the steam nozzle. If drier steam is required, additional separation may be obtained at the connection to the heating equipment; if super-heat is required, it may be accomplished by an externally, fuel-fired superheater. Fuel efficiencies are 80% for fuel oil and 78% for natural gas.

These units are positive in pressure to the gas exit only. If air pollution need not be considered, stub stacks may be used. Due to the draft effect of a tall stack, a multiple leaf damper with an automatic controller must be furnished to maintain design furnace conditions.

Although these steam generators are furnished completely packaged, they can be purchased knocked-down and erected on the job site.

These units are of low head room design and require a space approximately equal to the width, height and length of the generator to pull tubes for servicing. The unit should be set upon a 4-in. high, concrete, housekeeping pad.

Pressure parts must comply with applicable sections of the ASME Code for fired, high pressure, steam producing vessels. Water level control is achieved by an electric float switch that starts and stops the generator feed pump. Where refractory is furnished it should be of the super-duty class for long life. Drum insulation is generally made of 2-inch thick glass fiber, covered with a metal

jacket supported on spacers to prevent crushing the insulation. An 18-inch wide walk-way on the top of the unit should be integral with the jacket.

Control panel—A steel-enclosed control panel, containing all controls and connections for electrical power supply and steam pressure sensing lines for the control systems, is mounted on a frame at the front end of the generator.

Combustion controls—All manufacturers furnish electric positioning type, combustion controls that are adequate. If metering types are desired, they may be furnished as either electrical or air operated devices. If air operation is desired, an air compressor would be required. However, the electric positioning control does so well, it is this author's opinion that metering controls would be a waste of money.

A 3-pass, modified, scotch marine steam generator. Courtesy of Orr & Sembower Inc.

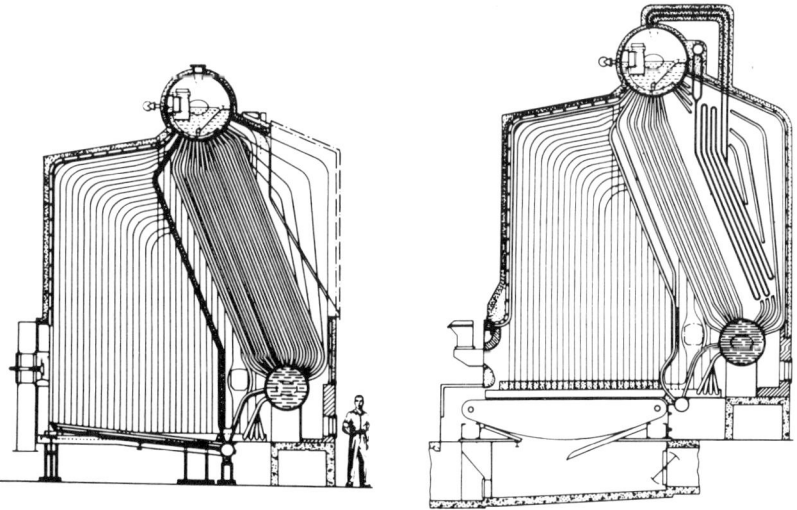

Side view through third pass, oil and gas fired (left), and side view through first pass, spreader stoker fired (right), integral furnace boilers. Courtesy of The Babcock & Wilcox Co.

Safety controls—It is the designer's responsibility to provide the maximum in safety. Design of safety controls and their arrangement should comply with all requirements of Factory Insurance Association (FIA), a standard in the industry. These include flame failure, complete fire programming and safety shutoffs for gas, fuel oil, steam overpressure, etc., and electrical interlocks.

A system of annunciating lights (one for each safety control) with an alarm should be furnished on the control panel to alert the operator to a trouble spot. There should be a pushbutton silencing switch for the horn, the light remaining on until the circuit is cleared. Color sequenced firing lights to indicate phases of the firing program should also be included in the design.

Fuel burners—These are steam generators designed to burn only the various grades of fuel oil, namely: Nos. 2, 3, 4, 5 and 6, natural and manufactured gas, or both oil and gas. All units are equipped with an electrically-driven combustion air fan to supply the necessary combustion air at the required static pressure to positively pressurize the furnace.

For the burning of gas, the unit is equipped with a gas ring for high-pressure gas furnished at approximately 1 psig.

The oil burner is of the air-assisted pressure atomizing type, a small air compressor being furnished integrally with the unit. Also furnished, but generally shipped separately as it must be located near the fuel tank, is the fuel oil pump. Mounted on the unit are electric and steam- or water-heated fuel oil heaters to raise heavy fuel oils to burning temperatures.

Tube cleaning—Due to the compactness of these steam generators it is not possible to install soot blowers. When burning oil, tubes must be cleaned with either a brush or vacuum cleaner. This is something of a draw-back, but the high velocity of the gases helps to sweep the tubes clean. When burning gas, very little cleaning is required. Fire tube baffles, sometimes called spinners, or any device placed in the fire tube entrance or exit must not be used as they trap soot in the tubes.

Field Erected Steam Generators

The fact that a gas- or oil-fired, packaged, steam generator costs less than a field erected one; can produce 200,000 lb steam per hour; occupies less space; operates at approximately 1000 psig; can be fitted with a super-heater, air heater and an economizer, as required, precludes the purchase of a field erected unit in this range and arrangement.

The burning of pulverized, crushed and stoker-sized coals, industrial waste or by-products such as bark, wood chips and bagasse, necessitates furnishing a field-erected steam generator with furnace and fuel burning devices arranged to handle these fuels.

In areas where these fuels are the most economical, field-erected designs are a necessity and, depending on size and requirements, may be arranged for underfeed stokers, spreader stokers, crushed coal burners and pulverized coal burners.

Due to the fact that there are many designs of steam generators available for the designers' selection, it is not possible to indicate all of them.

The varied integral furnace designs are typical. In these designs, all fuels may be burned, as they provide practically a universal arrangement. The units may be arranged for either indoor or outdoor installation. Some manufacturers design units with the steam drum directly over the mud drum, while others incline the steam drum forward. This design may be equipped with an air-heater or economizer as required.

Among the various integral furnaces are:

Type FF—Size range: steam output to 50,000 lb per hour, in increments of 2000 lb per hour or less. Design pressure: 160-600 psi. Operating pressure: 15 psi to safety valve setting. Steam temperature: saturation to 850F, superheat. Fuels: oil and gas, coal with all types of stokers. Furnace: water cooled, except floor, non-pressure type.

Type FP—Size range: steam output to 80,000 lb per hour in increments of 5000 lb. Design pressure: 160-825 psi. Operating pressure: 15 psi to safety valve setting. Steam temperature: saturation to 850F. Fuels: oil and gas, coal with chain grate, spreader or underfeed stokers. Furnace: water cooled, non-pressure type.

Type FH—Size range: steam output from 50,000-350,000 lb per hour in small increments. Design pressure: 160-1050 psi. Operating pressure: 15 psi to safety valve setting. Steam temperature: saturation to 910F. Fuels: pulverized coal, oil and gas, in combination or singly. Furnace: water cooled, non-pressure type.

Steam Plant Auxiliaries

Auxiliary equipment consists of equipment that steam plant designers consider essential or desirable, depending on the service requirements and economics of a given plant.

This equipment comprises the standby electric generator; blow-off tank and condensate cooler; flash tank desuperheater; steam pressure reducing valve; and steam traps.

Standby Electric Generator

Undoubtedly, the most important auxiliary to the steam plant is the standby electric generator, as demonstrated during power failures and "brownouts", particularly in the northeastern section of the U.S.

Plants served by overhead transmission lines are probably the most vulnerable to power interruption, and in these cases the standby electric generator can provide electric power for normal operation.

These electric generators may be operated by diesel engine, gas turbine or other type of internal combustion engine, using diesel fuel, natural gas, manufactured gas, bottled butane or propane.

Ranging from 30 kw to 800 kw, such engine generator units can run continuously day and night. Using natural gas, fuel consumption can be as little as 11,500 Btu per kw-hr; using diesel fuel, consumption may be as low as 0.55 lb per kw-hr.

The engine size specified should be sufficient to handle the electric load for the entire plant.

The author favors the use of diesel fuel because it is available practically everywhere. Although it may have a different name or designation in various parts of the world, it is actually No. 2 heating oil, a low viscosity distillate.

It is important that the designer make a complete survey of operating requirements, taking into consideration the past history of local outages, to arrive at a realistic fuel storage figure. Of course, if natural or manufactured gas is selected as the fuel, there is no storage problem. (Storage of diesel fuel usually comes under Municipal Fuel Oil Storage Code.)

Standby engine generators should be arranged to "come on line" automatically upon failure of the main power, and be able to reach operating speed within 14 seconds.

When a power failure occurs, all normal operating equipment should go on "Safety" automatically. Restart must be done manually by the plant operators.

The best location for the engine generator unit is inside the steam or hot water plant, where it is warm and dry. This location permits easy starting without the need for water jacket and lubricating oil heaters.

The unit should be mounted upon a heavy reinforced concrete base, and if isolation is required, pad or spring type isolators are generally sufficient. The required quantities of combustion and ventilation air indicated by the manufacturer must be provided, and this generally involves duct work, and possibly supply air fans.

9

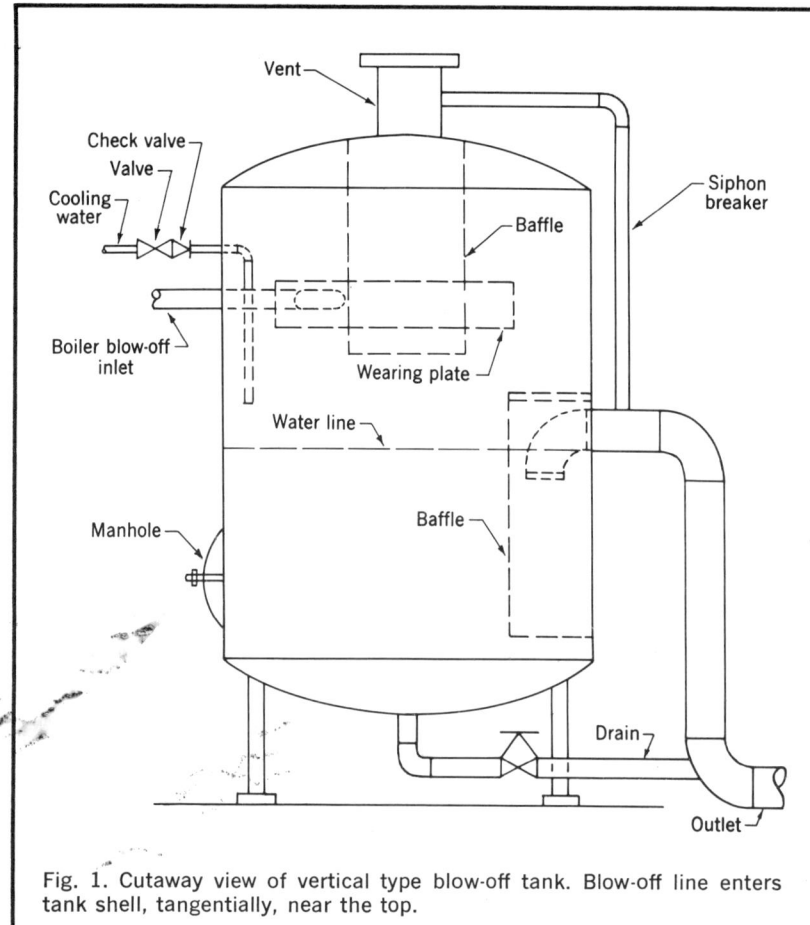

Fig. 1. Cutaway view of vertical type blow-off tank. Blow-off line enters tank shell, tangentially, near the top.

Outdoor air is used, with or without tempering, and if the unit is quite large, an outdoor heat exchanger, with fan, is used to control the temperature of the jacket water.

With this arrangement, ethylene glycol should be mixed with the jacket water to prevent freezing in cold areas.

Engines, dependent upon size, have one or two electric cranking motors, supplied with low voltage DC electric power from batteries maintained at their proper energy level by a trickle charger.

Engine generators can be furnished complete with circuit breaker, cranking panel, batteries with charger, automatic transfer switch and engine generator controls.

Blow-Off Tank

The mineral content of the water for a steam generator must be carefully controlled to prevent scale.

This control is accomplished by blow-down of a certain percentage of the water from the steam drum and mud drum into a blow-off tank, where it is condensed.

Basically, two types of blow-down tanks are used: horizontal, in which the blow-down line enters the shell at the top; and vertical, in which the blow-down line enters the shell tangentially, near the top. See Fig. 1. Both types work equally well.

The blow-down leaving the blow-off valve is a mixture of hot water and flash steam. A certain percentage of the hot

water flashes into steam as it reaches atmospheric conditions, but the larger portion remains as hot water. Steam is relieved to atmospher through a vent line that should be fitted with an exhaust head.

The tank should be of ASME code construction, all welded black carbon steel with shell and heads not less than 3/8 in. thickness.

The actual blow-off piping between generator and blow-off tank should never be less than Schedule 80 thickness (and heavier as required) black carbon steel or wrought iron, of welded construction.

The tank may be set on steel saddles, close to the floor, or elevated, as required.

Sizing the Blow-Off Tank

The blow-off tank must be sized in relation to the quantity of water to be handled. The first step in calculating the water volume and blow-off tank size is to determine the percentage of flash steam.

Assume that back pressure in the blow-down piping and system is 5 psig and steam pressure in the generator is 200 psig, then

$$\text{Flash steam \%} = \frac{SH - SL}{LH} \times 100$$

where

SH = sensible heat in condensate at 200 psig

SL = sensible heat in condensate at 5 psig

LH = latent heat of steam at 5 psig

Using standard steam table values:

$$\text{Flash steam \%} = \frac{361.91 - 196.16}{960} \times 100 = 17.25\%$$

Next, assume the following conditions:

Steam generator operating pressure: 200 psig

Steam drum: 4 ft dia × 16 ft long

Blow-down: 4-in. water column (at horizontal centerline of drum)

Blow-down tank pressure: 5 psig

Specific volume saturated water at 200 psig: 0.01847 cu ft per lb

Specific volume saturated water at 5 psig: 0.01683 cu ft per lb

Then:

Volume of blowdown water = 4 × 16 × 0.333 = 21.31 cu ft

Weight of blowdown water =

$$\frac{21.31}{0.01847} = 1150 \text{ lb}$$

Weight of water flashed to steam = 1150 × 17.25 = 198.37 lb (17.25% from previous calculation)

Net weight of blowdown water = 1150 − 198.37 = 941.63 lb

Holding volume, from bottom of tank to outlet = 941.63 × 0.01683 =15.84 cu ft

Since an equal volume is required for actual blowdown,

total water volume = 2 × 15.84 = 31.68 cu ft.

Since steam separating volume equals twice total water volume, total volume of tank is 2 × 31.68 = 63.36 cu ft.

Consider a tank with diameter of 3 ft 6 in. and a cross section of 9.61 sq ft.

Then,

$$\frac{63.36}{9.61 \times 1} = 6.6 \text{ or } 7 \text{ linear ft.}$$

Consequently, blowdown tank required is 3 ft 6 in. dia × 7 ft long.

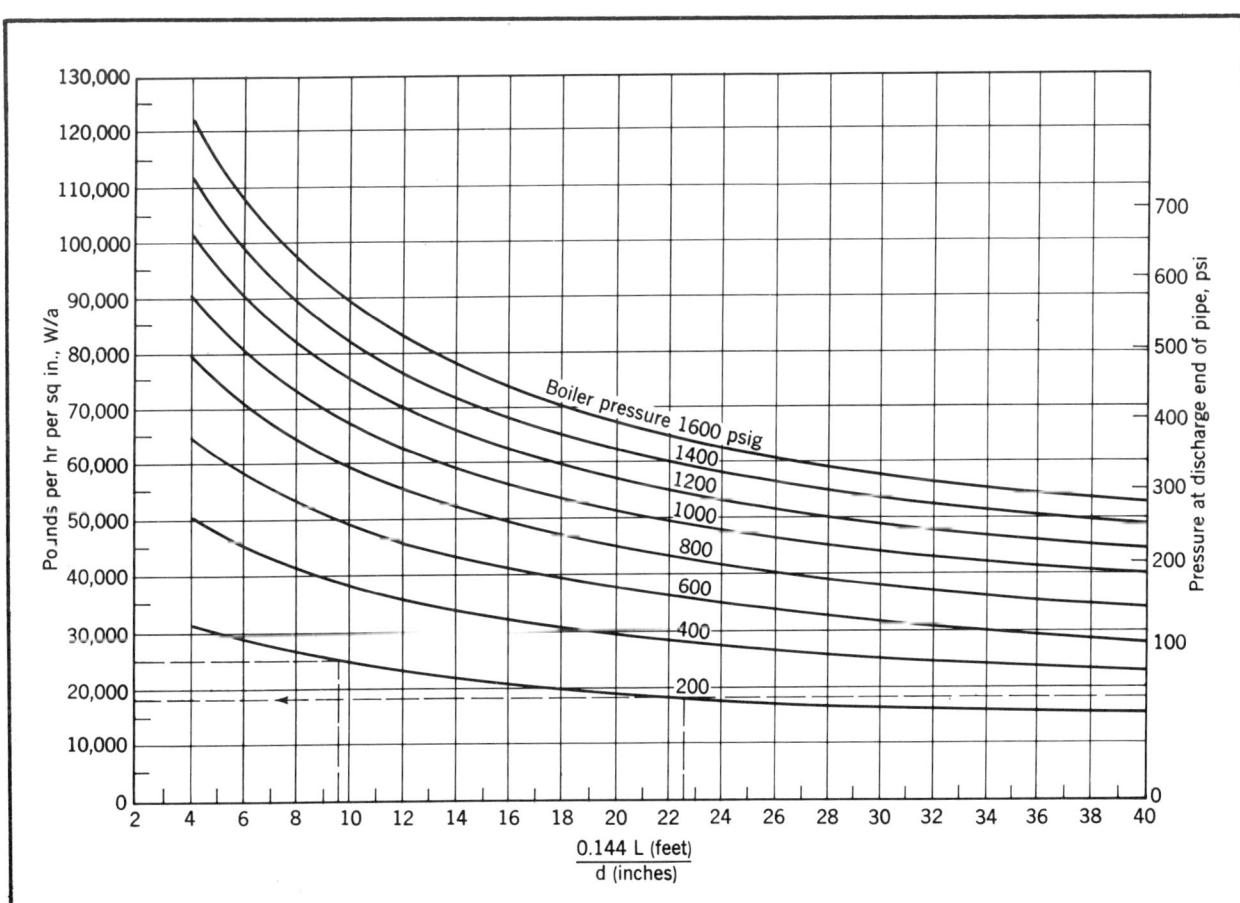

Fig. 2. Nomogram simplifies sizing of vent line for blow-off tank. After determining that equivalent length of blow-down line is 100 ft and inside diameter is 1.503 in, the formula $\frac{0.144 \, L(ft)}{d \, (in)}$ yields a factor of 9.58. Reading the nomogram from 9.58 vertically to the intersection with the 200 psig (operating pressure of boiler) curve and horizontally to the left hand scale shows that the water quantity is 25,500 lbs per hr per sq in of pipe area.

Table 1. Steam Load vs. Flash Tank Size

Condensate, lb per hr	Tank Dia, inches
0 - 500	6
500 - 1000	10
1000 - 1500	12
1500 - 2500	14
2500 - 3000	16
3000 - 4000	18

Sizing Vent Line

Assume that equivalent length of blowdown line is 100 ft and pipe is 2-in. double extra heavy.

Then inside diameter of pipe is 1.503 in. and internal area is 1.774 sq in.

Next, use the Nomogram in Fig. 2.

Step 1.

$$\frac{0.144 \times L}{d} = \frac{0.144 \times 100}{1.503} = 9.58$$

Step 2. Reading the Nomogram from 9.58 to intersection with 200 psig curve and directly left, shows the water quantity to be 25,500 lb per hr per sq in. of pipe area.

Therefore, flow through pipe of 1.774 sq in. internal area = 25,500 × 1.774 = 45,237 lb per hr.

Flash steam factor is 17.25% at 5 psig. 45,237 × 17.25 = 7800 lb steam per hr at 5 psig.

Using the steam pipe sizing chart for Schedule 40 pipe shows that a 5-in. pipe with a pressure drop of 3.3 lb per 100 ft will meet requirements.

Sizing Tank Discharge Line

Blowdown per hr = 45,237 lb
$\qquad\qquad\qquad\qquad$ (a)

Amount flashed to steam = 7800 lb $\quad$ (b)

Therefore, amount discharged from tank is a − b = 37,437 lb.

Weight of water at 5 psig = 7.9 lb per gal

$$\frac{37,437}{7.9 \times 60} = 79 \text{ or } 80 \text{ gpm}$$

Cameron Hydraulic Tables* show that a 4-in. extra strong steel pipe will easily handle a flow of 80 gpm with a head loss of 1 ft or 0.44 lb per 100 ft at 2.23 ft per sec.

Condensate Cooler

Since most municipalities require hot water effluent to be cooled to at least 140F before discharge into the sanitary system, some means of cooling must be used.

This cooling is achieved by a thermostatically controlled tank or condensate cooler that automatically injects sufficient cold water to achieve the required exit temperature.

The thermostat is set in the cooler outlet as close as possible to the cooler.

The amount of cold water required to cool the blow-down to 140F can be calculated as follows:

$$Gh = \frac{tm - tc}{th - tm}$$

where

Gh = gals hot water cooled by 1 gal cold water to yield a mixture at 140 F

tm = temp of mixture, 140 F

tc = temp of cold water, 60 F

th = temp of hot water, 212 F

Then

$$Gh = \frac{140 - 60}{212 - 140}$$
$$= 1.1 \text{ gal hot water}$$

*Published by Ingersoll-Rand Co., Cameron Pump Div., New York, N.Y.

Rate of blowdown (at 212 F) entering cooler is 132 gpm.
Cold water required is 132/1.1 = 120 gpm

To determine the time required for water blowdown for 4 in. of gage glass and the time during which cooling water flows — these figures are the same — the following example is given:

Blowdown rate is 45,237 lb per hr or 753 lb per min

Weight of water blowdown (4 in. gage glass) = 1150 lb

Weight of saturated water blowdown per min: 45,237/60 = 753 lb

Time required is 753/1150 = 0.654 min or 39.2 sec.

The designer will probably be surprised at the quantity of cooling water required, but this flow

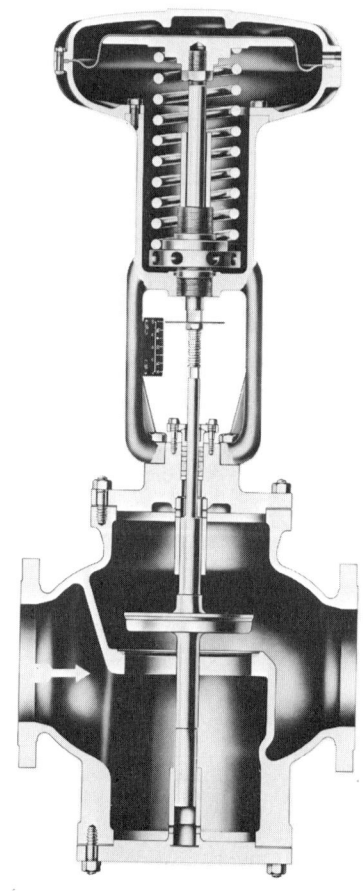

Fig. 3. A typical pressure reducing valve.

lasts little more than half a minute.

The condensate cooler should be sized so that the water holding section (below the outlet) contains at least 250 gal.

A 6-in dia pipe is adequate for the outlet connection.

The air in the tank displaced by the water must be vented to atmosphere, and a 2-in. line connected to the top of the tank may be run and connected to the blow-off tank vent.

Cooling water is controlled by a self-contained thermostatically operated valve, in which the thermal bulb is set in the outlet of the condensate cooler. The condensate cooler should be located as close as it is practical to the blow-off tank.

Flash Steam

Condensate drainage from medium and high pressure process equipment produces flash steam. The quantity varies according to the quantity of condensate and the pressures involved.

Flash steam is useful because, at the pressure it is produced, it retains its latent heat.

For example, a process unit uses 5000 lb steam per hr at 100 psig. The condensate in the trap is at 100 psig and the trap discharges into a flash tank connected to a heating main operating at 10 psig.

A certain percentage of this condensate will falsh into steam at 10 psig, at which pressure the latent heat of steam is 952.1 Btu per lb.

To calculate the percentage of condensate that will flash into steam, proceed as follows:

$$\% \text{ flash steam} = \frac{SH - SL}{LL} \times 100$$

where

SH = sensible heat in condensate at 100 psig

SL = sensible heat in condensate at 10 psig

Fig. 4. A steam pressure reducing station. Four remote pilots and one master pilot on center horizontal pipe operate four pneumatically controlled valves on top of two pipes.

LL = latent heat in steam at 10 psig

Therefore, flash steam % =

$$\frac{(308.8 - 208.4) \times 100}{952.1} = 10.5\%$$

$5000 \times 10.5\% = 525$ lb steam per hr

This means that if 5000 lb of condensate were allowed to flash down to an atmospheric receiver, the resulting loss would be 525 lb steam per hour or 525 x 952.1 Btu per hr, a total of approximately 500,000 Btu per hr.

Specifying the Flash Tank

Flash tanks may be constructed of black steel pipe to handle the operating pressure. All joints should be welded.

The tank should be mounted vertically with the steam outlet at the top and the water outlet 4-in. above the bottom. The condensate inlet should be 12-in. above the bottom of the tank.

The steam trap that drains the flash tank should be large enough to handle three times the normal load of condensate during cold start-up.

Table 1 shows the correlation

between steam load and flash tank diameter.

Pressure Reducing Valve

The purpose of the steam pressure reducing valve is to maintain a constant, predetermined terminal steam pressure by automatically replenishing the steam as it is used. A typical steam pressure reducing valve is shown in Fig. 3.

Steam pressure reducing valves are of two types: self-contained and air operated.

In operation, the self-contained valve senses down-stream steam pressure and a pilot line transfers this pressure to a steam operated pilot valve or directly to a diaphragm atop the pilot valve, exerting pressure on the valve stem and against an adjustable spring. The down-stream pressure is controlled by the tension of this spring.

The air controlled valve operates much in the same manner except that the control equipment senses down-stream steam pressure and sends an air impulse to a pilot valve, or directly to the valve diaphragm, to control the terminal steam pressure.

There are many variations of these steam pressure reducing valves to meet any required quantity and pressure of steam. See Fig. 4 showing a steam pressure reducing station.

Sizing the valve to pass the maximum and minimum quantity of steam is very important. Consequently, the designer should always confer with the valve manufacturer to obtain recommendations for the correct size and arrangement.

The Desuperheater

Superheated steam is dry saturated steam heated to a temperature above its saturation temperature to provide a more economical source of energy for the steam turbine.

To control the temperature of the superheater and to reduce the temperature of the turbine exhaust steam for process use or for space heating requires some means of lowering the superheat temperature.

This is the function of the desuperheater, and it achieves the cooling to the desired level by spraying water into the steam.

Heat released by the steam during the temperature reduction is picked up by the cooling water in three steps:

1. the cooling water temperature is raised to that of saturated water;
2. the heated water evaporates into steam;
3. this steam is heated to the required temperature at the desuperheater outlet.

One of the most popular desuperheaters is the variable orifice spray type. In smaller sized orifices, a metal ball provides turbulence and consequently an excellent mixing of spray water and steam. In larger sized orifices, turbulence is created by a plug that changes its position according to the steam flow to yield the desired mixing.

The spray water controller is proportional and maintains steam temperature within ±5F of the setting.

Turndown ratio is at least 50 to 1 and the final temperature may be sensed 20 ft downstream from the desuperheater. The final temperature may be as low as 10F above saturation.

Boiler feedwater may be used for cooling water and boiler feed pump pressure as the injection medium. If boiler feed pump pressure is too high, a water pressure reducing valve is used.

Where lower pressures and lower temperatures are desired, steam pressure is reduced by a conventional steam pressure reducing station and steam temperature by the desuperheater. Both items are usually sold as an integrated package with all controls.

Desuperheating and pressure reducing stations can be arranged to provide steam at several different pressure and temperature conditions.

Steam Traps

Steam traps are used to drain condensate from steam headers, mains, risers, steam separators, steam supply lines to steam turbines, heat exchangers and wherever else it is necessary to remove condensate.

The two types of steam traps generally used in steam plants are the inverted bucket and the thermodynamic. While each operates on a different principle, both use steam pressure as the motivating force.

In the inverted bucket type, a bucket is mechanically linked to the trap discharge valve. As the condensate flows into the trap, the bucket rises and closes the discharge valve; as the condensate continues to flow, it builds up air pressure under the bucket until a vent in the bucket relieves the pressure and bucket falls, allowing the discharge valve to open, at which point steam pressure forces condensate from the trap.

The thermodynamic steam trap has only one moving part, a disc that covers the discharge port and operates as a valve to control the exit of air and condensate from the trap. Condensate and/or air pressure lifts the disc off the discharge port seat and steam pressure forces the

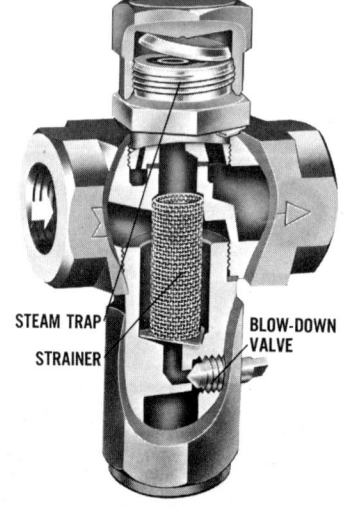

STEAM TRAP STRAINER　　　BLOW-DOWN VALVE

Fig. 5. Cutaway views of an impulse steam trap.

condensate and air through the discharge port.

Discharge continues until flashing condensate approaches steam temperature, when high velocity jets of flash steam move radially across the underside of the disc to reduce the pressure underneath the disc. Meanwhile, the steam builds up pressure in the control chamber above the disc and forces the disc against its seat, thus closing the trap. See Fig. 5.

Electric Motors

Electric motors are used in steam plants to drive pumps (boiler feed, feed water transfer, high temperature water circulation) and to drive fans (forced draft, induced draft and ventilation); air compressors, fuel pumps, etc.

For these services and practically all needs, the type of motor used is the squirrel cage.

Generally housed in drip-proof and weather-proof enclosures, these motors, regardless of size, may be started across-the-line. The only limits are those imposed by the power supply, the plant wiring system and the load.

Due to new and improved types of insulation that allow higher operating temperatures, today's motors are comparatively smaller for any given horsepower than the motors available 10-20 years ago.

Motor Starters and Circuit Breakers

Motor starters and circuit breakers should be combination units in a single enclosure. This arrangement reduces space and avoids the additional wiring used with separate units.

Starters used for steam plants should be across-the-line type, sized according to the motor they will handle.

The circuit breaker has only one function, to interrupt the electric current feeding the motor when an overload occurs and thus prevent damage to the motor.

Circuit breakers should be sized to handle 2.5 times the rated motor load.

Disconnect Switches

Fused disconnect switches should be provided in the feeder serving all electrical equipment.

Unfused disconnect switches should be provided at all remote, electrically operated equipment to protect the operator.

3

Feedwater Systems

The deaerating feedwater heater is a very necessary part of the steam generator feedwater system. Its two functions are: to remove oxygen and carbon dioxide from the feedwater and to heat the feedwater to the required temperature for injection into the steam generator.

These heaters generally operate at a pressure of 5-10 psig, but can operate at 50 psig, if a higher water temperature is required. The boiler feed pump passes the feedwater through single or multiple shell and tube heat exchangers to attain it.

The deaerating heater is designed to remove oxygen from the feedwater to a point where it contains no more than 0.005 cubic centimeters of oxygen per liter, generally written as 0.005 cc/L.

The carbon dioxide is easily released. The unit is furnished with a small valved vent that continuously emits a small plume of steam to the atmosphere, the oxygen and carbon dioxide leaving with the steam. Non-condensibles are released through an internal vent condenser of stainless steel construction. The tank used is generally an ASME Code tank designed for 30-50 psig.

The feedwater heater should be designed for the total plant capacity. With a properly sized surge tank, its storage effect need only be 5 minutes. This helps to reduce heater cost. Feedwater is heated economically by exhaust

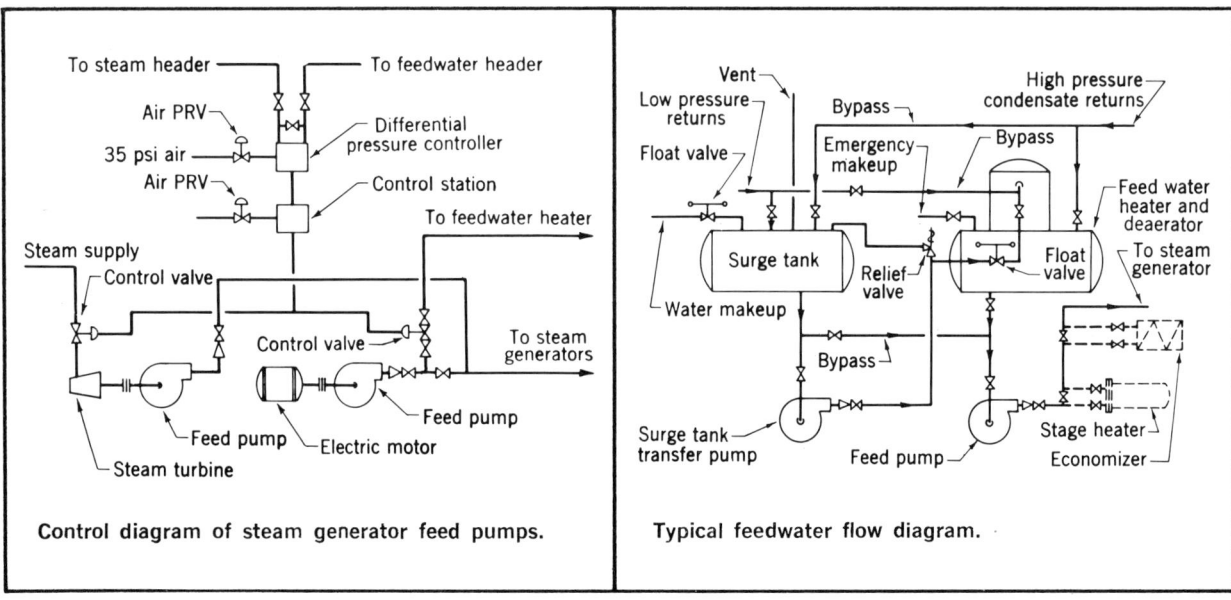

Control diagram of steam generator feed pumps.

Typical feedwater flow diagram.

16

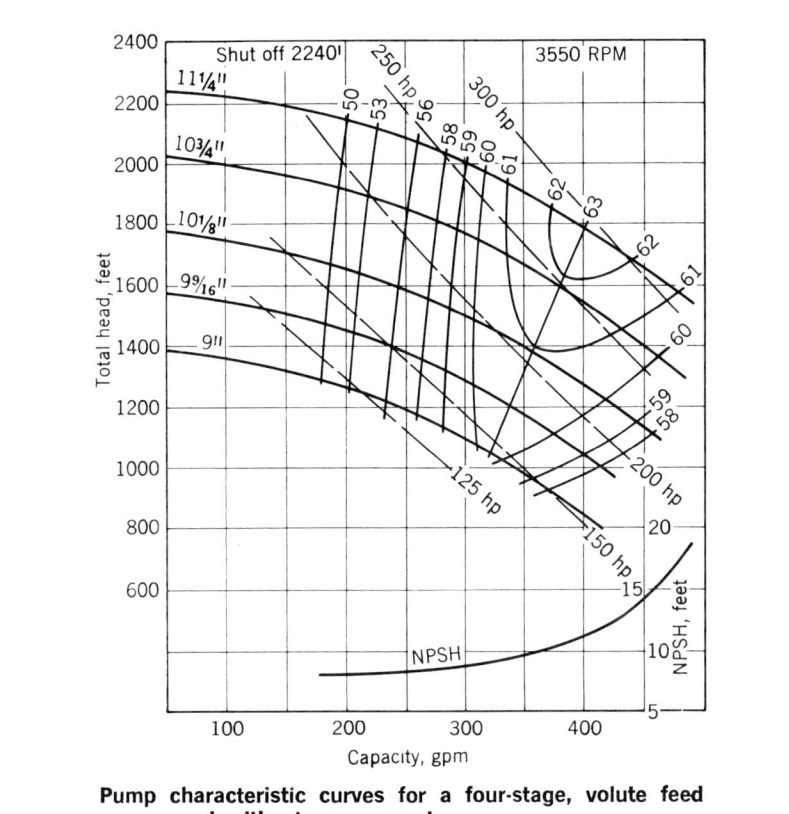

Pump characteristic curves for a four-stage, volute feed pump used with steam generators.

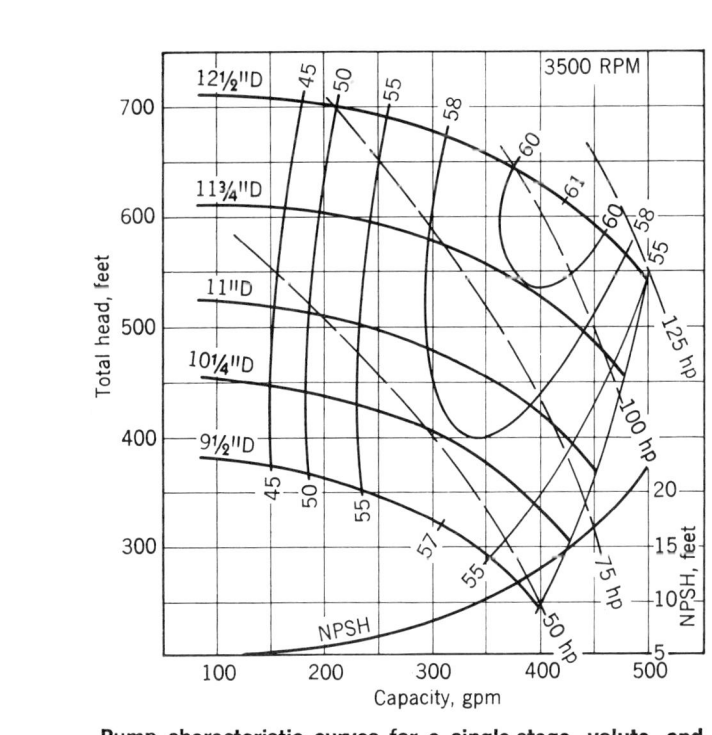

Pump characteristic curves for a single-stage, volute, end suction, steam generator feed pump.

steam from the steam generator feed and surge tank pump turbines, supplemented by live steam from a pressure reducing station.

Types of Feedwater Heaters

There are three types of deaerating feedwater heaters: tray or jet-tray, spray or atomizing, and atmospheric (which also employs a spray and an external heater).

The tray type heater and deaerator has multiple stainless steel trays, set one above the other, over which the returned condensate cascades downward, releasing the oxygen and carbon dioxide that passes through an internal vent condenser to the atmosphere. Water breaks up over the trays, and is heated by live steam from the generator injected into the heater.

Tray deaerators may be of the parallel downflow or counterflow designs. They may be furnished in vertical or horizontal designs to meet space conditions. Feedwater in the heater assumes the temperature of the steam.

The *spray type* unit differs from the tray design in that steam is utilized to break up the water, rather than trays. This is done by means of a variable or fixed orifice atomizer. Live steam is injected into the heater to bring the water up to temperature.

These heaters are equipped with overflow traps or valves, atmospheric relief valves (to maintain pressure in the heater), safety relief valves, high and low level alarm switches, thermometers and float control valves to maintain water level constant, condensate being pumped in through a float valve.

Spray and tray type deaerating heaters should be set at a height of at least 16 ft above the centerline of the boiler feed pumps to provide a positive suction head on the feed pump chamber to prevent flashing in the pump chamber. This is a good arrangement for, should the selected feed pump perform a bit critically on the suction side, the designer has a built-in safety factor.

Advantage of these two types of heaters is that they can operate

**A typical tray-type deaerating heater.
Courtesy of Cochrane Corp.**

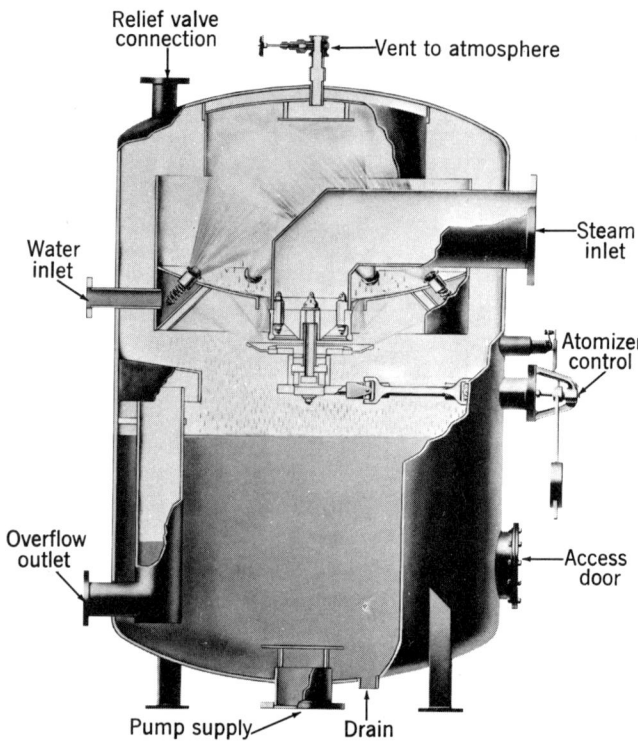

**A typical, atomizing-type, deaerating heater.
Courtesy of Cochrane Corp.**

at pressures up to 50 psig and furnish feedwater at that pressure and temperature without additional external water heaters.

The *atmospheric type* feedwater heater maintains the water in the receiver at atmospheric boiling point. The returning condensate is pumped through an external shell and tube heater, where the temperature is raised to 225F. It then enters the spray manifold in the heater and is flashed down to 212F and atmospheric pressure, releasing oxygen and carbon dioxide from the water. The gases leave the heater through a vent condenser.

If higher water temperatures are required or desired, additional shell and tube heaters located in the boiler feed pump discharge piping are required to raise the water temperature to higher levels.

Surge Tank

The surge tank is the collecting tank for all low temperature con-

densate returns. Also known as a receiver, it is vented to the atmosphere to allow water to enter. It is fitted with a gage glass and a float operated, makeup water valve that helps to maintain a stable water level in order to furnish a constant, positive suction head for the surge tank pumps. These tanks are usually constructed of $\frac{3}{8}$-inch thick, black steel plate with welded seams. The surge tank and feedwater heater generally share an elevated platform, or mezzanine, in the steam plant.

The surge tank should be sized to provide storage for at least 15 minutes. When called for, this storage effect should use only three-quarters of the tank capacity. At this level, an over-flow line should be provided to take care of any unusual surges. The line should discharge to the condensate cooler that also handles hot blowdown from steam generators.

Feed Pumps

Function of the feed pumps is to continuously supply the steam

generator with heated and deaerated feedwater. These pumps are generally of the horizontally or vertically split volute end suction types, driven either by electric motor or steam turbine.

Steam generator feed service is of the toughest for pumps because they must operate continuously under elevated temperatures and pressures. The pumps must be of substantial design and workmanship and all materials used in their construction should be the finest obtainable. Pump shaft glands may be sealed with packing or balanced mechanical seals. Speeds range from 1750-4000 rpm.

Depending on pressures, pump casings are made of either cast iron or cast steel. Impellers may be made of cast iron, steel, bronze or stainless steel, as required. Pump shafts should be of substantial thickness tempered steel, riding on ball bearings, immersed in oil, having a water jacket to control oil temperature. Shaft seal areas are also water cooled.

To be efficient and avoid prob-

Heat Balance Example

To show the quantity of steam necessary to supply a feedwater heater and the economics of using exhaust steam to heat the feedwater, it is necessary to make a heat balance. The plant steam load is 40,000 lb per hour at 135 psig steam generator drum pressure.

Power requirements are:

One feed pump, electrically driven	20 hp
One feed pump, steam turbine driven	20 hp
One surge tank pump, electrically driven	10 hp
One surge tank pump, steam turbine driven	10 hp
Two turbines	30 hp
Turbine water rate, 55 lb per bhp per hour $\times$ 30 bhp = 1650 lb steam	
Back pressure on turbines	5 psig
Latent heat of steam at 5 psig	961 Btu per lb
Cold water makeup at 40F	2000 lb per hour
Condensate returning at 180F	38,000 lb per hour
Maintain feedwater at 226F	
Return condensate temperature differential, 226 − 180F	46F
Makup water differential, 226 − 40F	186F

$$\frac{38,000 \times 46}{961} = 1820 \text{ lb per hour}$$

$$\frac{2000 \times 186}{961} = 387 \text{ lb per hour}$$

1820 + 387 = 2207 lb steam per hour required to heat feedwater

Steam required to heat feedwater	2207
− Exhaust steam from turbines at 5 psig	1650
Total live steam makeup, lb per hour	557

Approximately 75% of the steam required to heat the feedwater is furnished at no additional cost by the steam turbine. The feedwater heater should be fitted wilth a steam pressure reducing station capable of furnishing the total amount of steam necessary to heat all the feedwater, should the electric pumps be operating.

Total Discharge Head Calculation

To arrive at the required discharge head of a feed pump to serve a steam generator operating at 230 psig, the pressures required throughout the system must be added. Pump suction is flooded to provide adequate Net Positive Suction Head (NPSH), as required by the pump.

Required pressures are:

Steam generator steam pressure	230 psig
Pressure drop through pump control valve	10 psig
Pressure drop through feedwater regulator	25 psig
Hydrostatic head (from pump to top of generator)	15 psig
Losses through pipe, valves and fittings	10 psig
Subtotal	290 psig
+ 10% factor for wear	29 psig
Total discharge head required	319 psig

Note: In the above example, all water is heated in a deaerating feedwater heater. If multistage heaters and an economizer are used (they would be in the pump discharge before the steam generator), their pressure drops must be added to the total discharge head of the pump.

lems of cavitation or recirculation, a pump should be sized so that it will be fully loaded, or nearly so, under all operating conditions.

Should conditions be such that for certain operating periods the load will be reduced, an additional, smaller feed pump should be selected for use during such periods.

Drives

Feed pumps are driven by an electric motor or steam turbine. These two types of drives furnish complete diversity for pump operation. The electric motor is the less expensive of the two. It allows the pump to be driven when exhaust steam in large quantities is not required, thus promoting plant heat balance.

As the motor operates at a constant rpm, an automatic control valve is necessary in the pump discharge to control flow to the steam generator.

These motors may be of the squirrel-cage, induction type and of open, splash-proof construction. They may be started across-the-line at full voltage and should have a combination across-the-line motor starter and circuit breaker equipped with auxiliary contacts, reduced voltage transformer, start-stop pushbutton and operating lights.

Design of the steam turbine to drive the feed pump has been developed through the years. It is sometimes known as a mechanical drive turbine. Actually, this type is a single-stage, impulse turbine, having a bladed wheel or milled buckets. The unit operates at steam generator pressure is non-condensing and generally furnishes steam to heat and feedwater at 5 to 10 psig steam pressure.

The unit is automatically controlled by a valve in the steam supply that changes with the speed and controls feedwater supply to the steam generator.

Advantage of using the turbine is that the steam used to pump the feedwater may also be used to heat it. This naturally improves the plant heat balance.

Feed Pump Control

Whether the feed pump is electrically- or steam-turbine-driven, it must be controlled to meet varying flow conditions. The electric pump is controlled by an automatic valve in its discharge. The steam turbine pump is controlled by an automatic valve in its steam supply.

A pressure controller senses feed line pressure and sends an air impulse to the control station that, in turn, loads or unloads the diaphragms of the automatic valves at the pumps. These, in turn, maintain the feed line pressure constant.

Surge Tank Pump Control

Control of surge tank pumps presents no problems because their discharge head is quite low. They are easily controlled by inserting a relief valve in the pump header and returning the relief line to the surge tank. For this service the pump should have a steadily rising head-capacity characteristic.

4

Principles of High Temperature Water Systems

High temperature water plants for process heat and space heating have been designed and constructed in the United States for some fifteen years.

And although the design principles for both these applications are similar, the engineer and contractor will be called upon more frequently to design and build the comparatively larger capacity space heating type plant. For this reason, the material presented here will be confined to the space heating category.

What Is High Temperature Water?

High temperature water (HTW) is hot water above 300F maintained at a pressure higher than its saturation pressure so that it remains stable and does not flash.

Medium temperature water is hot water between 250F and 300F; low temperature water is water below 250F.

How HTW Is Generated

HTW is usually heated in generators constructed of steel tubes and headers—no drums are used. Another arrangement, where a large amount of steam is required in addition to HTW, uses a cascade heater in conjunction with a steam generator.

Smaller amounts of steam can also be generated (in addition to HTW) by circulating the HTW through an unfired steam generator.

Pressurization of HTW systems can be done with steam or nitrogen. Steam works very well but requires larger expansion drums and more headroom than the more economical and efficient nitrogen. See Fig. 1.

Conventional fuels are used and the HTW plant operation is simple and automatic.

Advantages, Limitations, Costs

The two main features of HTW are:

Safety—it is the safest heat transmitting medium for the temperature and pressure ranges where it is used. Unlike steam, which contains latent heat, pressurized hot water contains only sensible heat and, when released to atmospheric pressure, it immediately flashes into a wet

"sub-cooled" · highly saturated water vapor. At a distance of two to three feet from the point of emergency HTW drops to 120F to 130F which does not cause burns if a pipeline fractures. On the other hand, high or low pressure steam upon contact with human flesh will cause a severe burn.

Simplicity—HTW is an easy medium to work with, and once the designer is alerted to the few pitfalls, he will enjoy working with it.

The HTW system lends itself best to large multiple building complexes, such as airports, shopping centers, industrial parks, colleges, satellite cities, etc.

Where heat is furnished from a central plant and distributed underground to the various structures, the actual cost of a central HTW plant is about the same as for steam, but the transmission lines for a HTW system are smaller, and therefore less expensive than for steam. The use of smaller diameter lines is due to the higher thermal capacity of water relative to steam. See Tables 1 and 2.

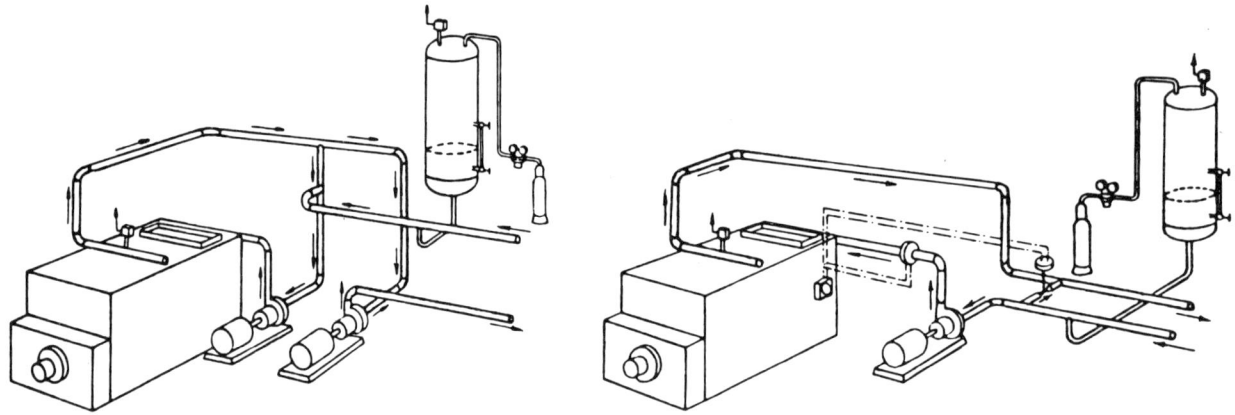

A. SEPARATE PUMPING SYSTEM. NITROGEN PRESSURIZED   B. INERT GAS PRESSURIZED SYSTEM.-COMBINED PUMPING

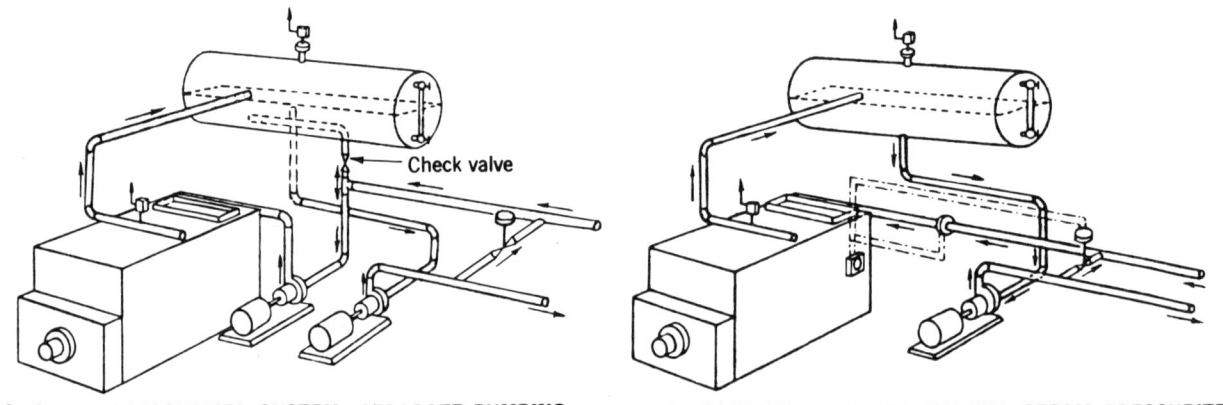

C. STEAM PRESSURIZED SYSTEM.-SEPARATE PUMPING   D. COMBINED PUMPING SYSTEM. STEAM PRESSURIZED

Fig. 1. Various methods of pressurization of HTW systems.

HTW piping may follow the contour of the site, and only manual vent valves are required at high points.

Pipe loops take care of pipe line expansion, which eliminates most of the costly access chambers required for expansion joints in steam systems.

In a HTW system there are no steam pressure reducing stations, steam traps, flash tanks or condensate pumps, which (in steam systems) are items of first cost and then become a maintenance cost.

Because a HTW system is a sealed system, pipeline corrosion is practically non-existent and maintenance is very low. There is no condensate return line corrosion, a major problem in the steam heating system.

Feed water treatment is necessary, but minimal, due to the sealed system.

In the HTW system, heat losses, other than the usual ones through insulation, are practically non-existent. However, in the steam system, these losses are considerable, for example:

1. flash tanks are vented to atmosphere;

2. feed water is continuously vented—to remove CO_2 and O_2;

Table 1. Comparison of Heat Content of Water and Steam

Absolute pressure, psi	Saturated temp, F	Return temp, F	Density, lb/cu ft		Heat content, Btu/cu ft		Heat content ratio
			Water	Steam	Water	Steam	Water to steam
14.7	212	180	59.8	0.0373	1923	37.4	51.5/1
29.8	250	200	58.9	0.0724	2972	72.1	41.2/1
67.0	300	200	57.3	0.1547	5820	156.7	37.2/1
134.6	350	200	55.6	0.299	8550	306.5	27.9/1
247.3	400	200	53.5	0.537	11090	555	20.0/1
343.7	430	200	52.4	0.741	12570	768	16.4/1
566.1	480	200	50	1.222	14820	1268	11.7/1

3. defective traps allow passing of steam;

4. condensate flashes down to atmospheric temperature and pressure and, particularly in large systems, must be heated before being returned to the feed water heater;

5. during blow-down of steam generator;

6. leakage at valve and pump glands . . . such leaks cannot be tolerated in a HTW system.

These steam generator losses (losses from flash steam alone have been computed to be more than 15%) mean more fuel, more feed water make-up, additional chemicals for feed water treatment and for pre-treatment of make-up water . . . all translated into higher operating costs.

On the other hand, HTW cannot compete with steam or low temperature water in the single large or small structure where distribution lines are short, because the HTW heating plant cost is much higher.

For instance, if low temperature water is generated, it can go directly to the user without the additional heat exchanger, pumps and transmission piping and produce the same end result at lower cost than HTW.

Heat Transfer

HTW is pumped through pipe lines from the central plant to the various buildings of the complex it serves. At these points the HTW is passed through heat exchangers, generating hot water in secondary systems at lower temperatures for distribution directly to the heat users.

By using separate pumps and three-way valves, HTW can also serve heating coils in air conditioning systems and may be used directly in large space heaters in high ceilinged factories.

DESIGN BASICS

Economical design of a HTW system for space heating is dependent upon several factors:

Table 2. Heat Storage Comparison of Piping in Equivalent Sized Steam and HTW Systems

Quantities	150,000,000 Btuh Output	
	Steam Load	HTW Load
Operating pressure, psia	135	290
Discharge temp, F	350	414
Return temp, F	180	250
Enthalpy at discharge, Btu/lb	1192.30	390.46
Enthalpy at condensate, Btu/lb	321.63	218.48
Enthalpy difference, Btu/lb	870.7	171.98
Flow rate, lb/hr	172,300	872,000
Density of fluid, lb/cu ft	0.30	53.0
Volume, cu ft/sec	159.5	4.58
Design velocity, ft/sec	112.5	6.1
Required cross section, sq ft	1.42	0.753
Equivalent extra heavy pipe size, in.	18	12
Pipe volume per 5000 ft pipe length, cu ft	7,100	3,750
Available heat, Btu/cu ft	261	9,110
Heat storage per 5000 ft of pipe, Btu	1,852,000	34,200,000
Time required to absorb stored heat	44.5(sec)	13.7(min)

1. It is of the utmost importance to employ the highest possible water temperature differential. Limits today indicate an economical water leaving temperature of 450F and a return water temperature of 200F. This means that at maximum load condition, each pound of water sent out will contain 262 Btu. Pressures, temperatures and heat content of water are given in Table 3.

(This is a far cry from low temperature water heating systems sending out hot water at 200F and returning it at 180F with a resultant transmission of only 20 Btu per pound of water.)

2. Proper sizing of valves is also important. A 300 psi cast steel valve will handle 650 psig working pressure at 450F in pipes larger than 2-in diameter. A 600 psi valve will handle pressures up to 1300 psig at 450F in pipes under 2-in diameter.

Since standard design valves for the HTW generator are rated for a temperature of 450F and a working pressure of 500 psig, and heat exchangers and pumps to handle this pressure and temperature range are readily available, the designer should work with these figures whenever possible.

3. Pipe and fittings must be welded. All pipe sizes up to and including 12-in schedule 40 pipe can be used. Pressure and temperature limitations for 12-in schedule 40 pipe are 640 psig and 650F. This 12-in diameter pipe can handle 1,000,000 lbs water per hour, or 262,000,000 Btuh with a pressure drop of 0.58 ft per 100 ft and a velocity of approximately 6 ft per second. A steam line handling the same number of Btu at 125 psig and a pressure drop of 1 lb per 100 ft, comparable figures for steam distribution, would have a diameter of 18-in.

Table 3. Selected Thermal Properties of Water and Steam*

Temp., Deg F	Saturation Pressure Psia	Saturation Pressure Psig	Saturated Water Specific Volume, cu ft/lb	Saturated Water Density, lb/cu ft	Saturated Water Specific Heat, Btu/lb/deg	Saturated Water Enthalpy Btu/lb	Saturated Water Enthalpy Btu/cu ft	Heat of Vaporization, Btu/lb	Saturated Steam Specific Volume, cu ft/lb	Saturated Steam Enthalpy Btu/lb	Saturated Steam Enthalpy Btu/cu ft	Temp., Deg C
32	0.089	—	0.01602	62.422	1.01	0.00	0	1075.8	3306	1075.8	0.3	0.0
50	0.178	—	0.01603	62.383	1.00	18.07	1,127	1065.6	1703.2	1083.7	0.6	10.0
80	0.507	—	0.01608	62.189	1.00	48.02	2,986	1048.6	633.1	1096.6	1.7	26.7
100	0.949	—	0.01613	61.996	1.00	67.97	4,214	1037.2	350.4	1105.2	3.2	37.8
110	1.275	—	0.01617	61.843	1.00	77.94	4,820	1031.6	265.4	1109.5	4.2	43.3
120	1.692	—	0.01620	61.728	1.00	87.92	5,427	1025.8	203.27	1113.7	5.5	48.9
130	2.223	—	0.01625	61.538	1.00	97.90	6,025	1020.0	157.34	1117.9	7.1	54.5
140	2.889	—	0.01629	61.387	1.00	107.89	6,623	1014.1	123.01	1122.0	9.1	60.0
150	3.718	—	0.01634	61.200	1.00	117.89	7,215	1008.2	97.07	1126.1	11.6	65.6
160	4.741	—	0.01639	61.013	1.00	127.89	7,803	1002.3	77.29	1130.2	14.6	71.1
170	5.992	—	0.01645	60.790	1.00	137.90	8,383	996.3	62.06	1134.2	18.3	76.7
180	7.510	—	0.01651	60.569	1.00	147.92	8,959	990.2	50.23	1138.1	22.7	82.2
190	9.339	—	0.01657	60.350	1.00	157.95	9,532	984.1	40.96	1142.0	27.9	87.8
200	11.526	—	0.01663	60.132	1.01	167.99	10,102	977.9	33.64	1145.9	34.1	93.4
210	14,123	—	0.01670	59.880	1.01	178.05	10,662	971.6	27.82	1149.7	41.3	98.9
212	14.696	0.000	0.01672	59.809	1.01	180.07	10,769	970.3	26.80	1150.4	42.9	100.0
220	17,186	2.490	0.01677	59.630	1.01	188.13	11,218	965.2	23.15	1153.4	49.8	104.5
230	20.780	6.084	0.01684	59.382	1.01	198.23	11,771	958.8	19.382	1157.0	59.7	110.0
240	24.969	10.273	0.01692	59.102	1.01	208.34	12,313	952.2	16.323	1160.5	71.1	115.6
250	29.825	15.129	0.01700	58.824	1.02	216.48	12,734	945.5	13.821	1164.0	84.2	121.1
260	35.429	20.733	0.01709	58.514	1.02	228.64	13,379	938.7	11.763	1167.3	99.2	126.7
270	41.858	27.162	0.01717	58.241	1.02	238.84	13,910	931.8	10.061	1170.6	116.3	132.2
280	49.203	34.507	0.01726	57.937	1.02	249.06	14,430	924.7	8.645	1173.8	135.8	137.8
290	57.556	42.860	0.01735	57.637	1.02	259.31	14,946	917.5	7.461	1176.8	157.7	143.4
300	67.013	52.317	0.01745	57.307	1.03	269.59	15,449	910.1	6.466	1179.7	182.5	148.9
310	77.68	62.98	0.01755	56.980	1.03	279.92	15,950	902.6	5.626	1182.5	210.2	154.5
320	89.66	74.96	0.01765	56.657	1.04	290.28	16,446	894.9	4.914	1185.2	241.2	160.0
330	103.06	88.36	0.01776	56.306	1.04	300.68	16,930	887.0	4.307	1187.7	275.8	165.6
340	118.01	103.31	0.01787	55.960	1.05	311.13	17,411	879.0	.788	1190.1	314.2	171.1
350	134.63	119.93	0.01799	55.586	1.05	321.63	17,878	870.7	3.342	1192.3	356.8	176.7
360	153.04	138.34	0.01811	55.218	1.05	332.18	18,342	862.2	2.957	1194.4	403.9	182.2
370	173.37	158.67	0.01823	54.855	1.05	342.79	18,804	853.5	2.625	1196.3	455.7	187.8
380	195.77	181.07	0.01836	54.466	1.06	353.45	19,251	844.6	2.335	1198.1	513.1	193.4
390	220.37	205.67	0.01850	54.054	1.07	364.17	19,685	835.4	2.0836	1199.6	575.7	198.9
400	247.31	232.61	0.01864	53.648	1.09	374.97	20,116	826.0	1.8633	1201.0	644.6	204.5
410	276.75	262.05	0.01878	53.248	1.09	385.83	20,545	816.3	1.6700	1202.1	719.8	210.0
420	308.83	294.13	0.01894	52.798	1.10	396.77	20,949	806.3	1.5000	1203.1	802.1	215.6
430	343.72	329.02	0.01910	52.356	1.11	407.79	21,350	796.0	1.3499	1203.8	891.8	221.1
440	381.59	366.89	0.01926	51.921	1.12	418.90	21,750	785.4	1,2171	1204.3	989.4	226.7
450	422.6	407.9	0.0194	51.55	1.12	430.1	22,170	774.5	1.0993	1204.6	1095.8	232.2
460	466.9	452.2	0.0196	51.02	1.14	441.4	22,520	763.2	0.9944	1204.6	1211.3	237.8
470	514.7	500.0	0.0198	50.51	1.15	452.8	22,870	751.5	0.9009	1204.3	1,337	243.4
480	566.1	551.4	0.0200	50.00	1.16	464.4	23,220	739.4	0.8172	1203.7	1,473	248.9
490	621.4	606.7	0.0202	49.51	1.17	476.0	23,570	726.8	0.7423	1202.8	1,620	254.5
500	680.8	666.1	0.0204	49.02	1.19	487.8	23,910	713.9	0.6749	1201.7	1,780	260.0
520	812.4	797.7	0.0209	47.85	1.22	511.9	24,494	686.4	0.5594	1198.2	2,142	271.1
540	962.5	947.8	0.0215	46.51	1.26	536.6	24,960	656.6	0.4649	1193.2	2,567	282.2
560	1133.1	1118.4	0.0221	45.25	1.31	562.2	25,440	624.2	0.3868	1186.4	3,067	293.4
580	1325.8	1311.1	0.0228	43.86	1.37	588.5	25,830	588.4	0.3217	1177.3	3,660	304.5
600	1542.9	1528.2	0.0236	42.37	1.45	617.0	26,140	548.5	0.2668	1165.5	4,368	315.6
620	1786.6	1771.9	0.0247	40.49	1.53	646.7	26,180	503.6	0.2201	1150.3	5,226	326.7
640	2059.7	2045.0	0.0260	38.46	1.66	678.6	26,100	452.0	0.1798	1130.5	6,288	337.8
660	2365.4	2350.7	0.0278	35.97	1.93	714.2	25,690	390.2	0.1442	1104.4	7,659	348.9
680	2708.1	2693.4	0.0305	32.79	2.47	757.3	24,830	309.9	0.1115	1067.2	9,571	360.0
705.4	3206.2	3191.5	0.0503	19.88	—	902.7	17,950	0	0.0503	902.7	17,946	374.1

*Data in columns 2, 4, 7, 9, 10, and 11 are taken from "Thermodynamic Properties of Steam" by Joseph H. Keenan and Frederick, G. Keyes, copyright 1937 by the authors and published by John Wiley & Sons, Inc., New York. All the remaining data were calculated from the data of Keenan and Keyes.

Minimum Safe Water Return Temperatures

As previously stated, it is of the utmost economic importance to design the system with the largest possible temperature differential so that each pound of water will contain the maximum number of Btu. However, there are two conditions that the designer must be aware of:

1. the minimum return temperature must be high enough to **prevent thermal shock** and

2. return temperatures must be safely above dew-point to prevent corrosion of steel tubes, headers and plate in the generators due to sulfur liberated from the fuel.

For instance, bituminous coals and No. 6, No. 5 and No. 4 blend fuel oils liberate sulfur on firing, whereas natural gas and No. 2 fuel oil do not.

Table 4. indicates the minimum safe water return temperature when burning various types of fuels with corresponding types of firing equipment.

Table 4. Minimum Safe Water Return Temperatures

Percent Sulfur* by Weight	Type of fuel*			
	Bituminous Coal			Fuel Oil No. 6
	Chain grate or underfeed	Spreader	Pulverized	
	Safe Water Temperature, F			
2	208	173	158	190
2½	226	183	167	207
3	239	193	177	223
3½	250	202	185	234
4	260	210	193	244
4½	268	218	200	254
5	276	224	207	258
5½	284	230	212	284

* Natural gas contains no sulfur and minimum recommended generator water inlet temperature is 160 F.

Most states, at this time, limit sulfur to 1% and are reducing this figure yearly.

No. 2 fuel oil contains no sulfur and minimum recommended generator water inlet temperature is 180 F.

5

High Temperature Water Generators and Auxiliaries

Basically, the controlled circulation HTW generator is similar to the large industrial type steam generator. For example, in both cases the furnace is surrounded by a radiant water tube surface; a convection section handles the hot combustion gases; and firing is done with conventional fuels. Insulation, refractory wall, roof and floor construction is also similar to steam generators.

But here all similarity ends— the HTW generator has no drums; its tubes are of small diameter, rolled and sealed or strength welded into headers; and it must be pump circulated by an individual pump or system pump, depending upon system design.

If the HTW generator is fired by oil or coal, the soot blowing system is of the air-puff type, and if plant air is not available, an air compressor and receiver must be furnished.

Soot blowers are not necessary when firing with gas because no carbon is deposited upon the tubes.

To provide low head-room design, the convection section is set behind the furnace section. Although this means a larger unit, it is preferable to a unit requiring high headroom.

Combustion chamber heat release, heat transfer per sq ft of surface and efficiency are the same as for a steam generator of comparable size.

Circulation, in some designs, is controlled by orificing the tubes, and in others by balancing the circuits.

Flow balance must be maintained or hot spots will be formed and cause steam to be generated in the tubes.

Average size HTW units are completely packaged (as are steam generators) and may be transported by rail or trailer. Larger sized units must be erected at the job site.

HTW generators are designed to efficiently produce only high temperature water. Although steam generators are primarily designed to produce steam only, some have been adapted to produce HTW, by changing inlet and outlet piping connections and inserting baffles and water jets in the drums to help promote water circulation.

Selecting HTW Generators

The size of the HTW generator depends upon the same load factors as the steam generator.

Using a large shopping center as an example, these load factors would be:
a. heating load
b. cooling load
c. in-between seasons load (heating and cooling)
d. domestic hot water load
e. steam load
f. night load (heating or cooling, domestic hot water and steam).

Obviously, the load will vary and, consequently, generator capacity must be elastic to meet any situation efficiently.

Since oil burners have an average turndown of 5 to 1, and gas burners 10 to 1, a 60,000,000 Btu unit, for instance, can turn down to 12,000,000 Btuh when oil fired, and to 6,000,000 Btuh when gas fired. And, because in most cases each generator has a single burner or one stoker and stoker control, there is no problem of controlling the output.

However, if oil is fired and the load drops to 10,000,000 Btuh,

26

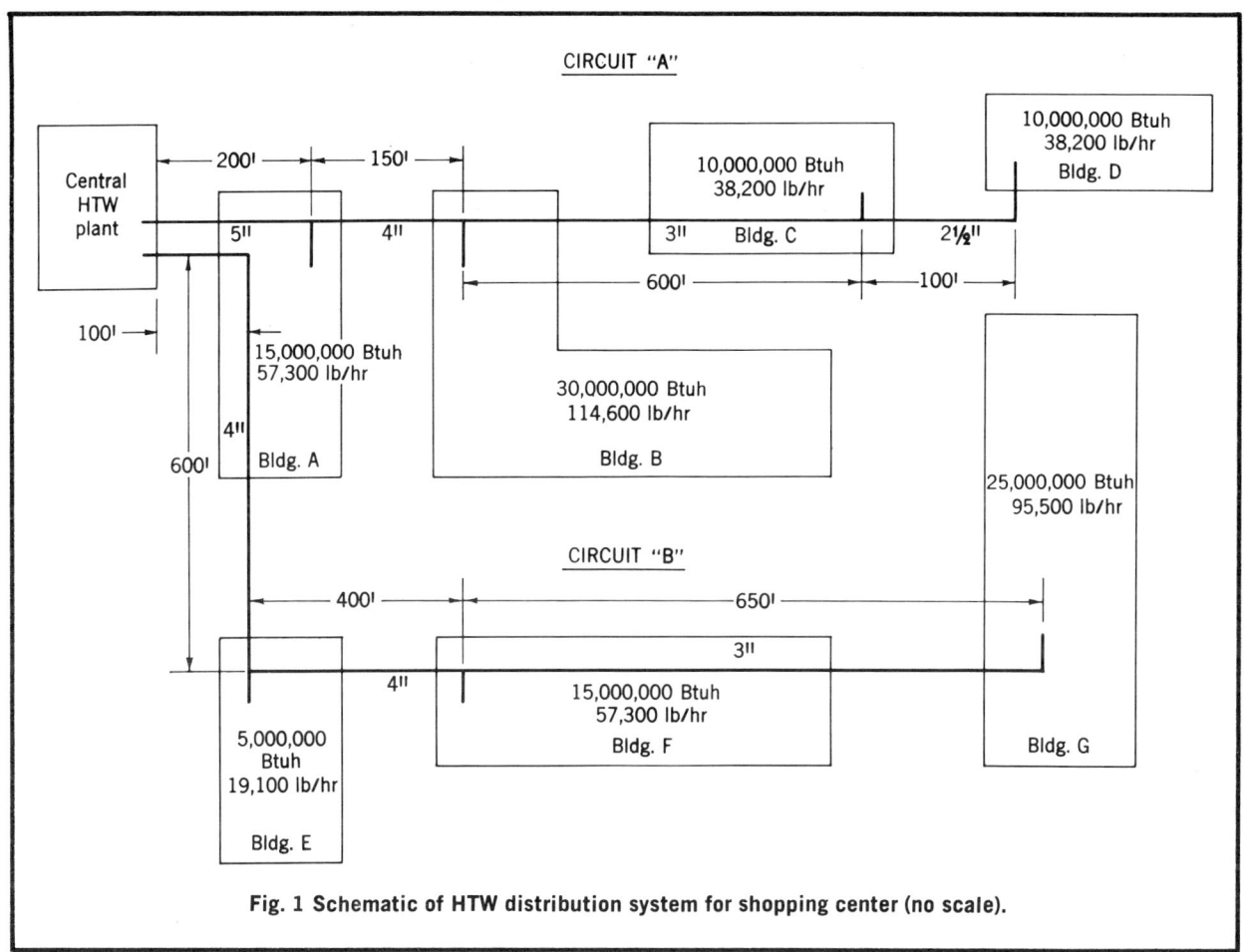

Fig. 1 Schematic of HTW distribution system for shopping center (no scale).

control will be lost and the burner will automatically go into a continuous short re-cycling operation which is not desirable. In such a case the only recourse is to try to control the load manually.

Therefore, to properly meet various loads, several different size generators must be installed, keeping in mind that HTW generators are efficient down to 25-30% of load.

A method of sizing HTW generators is shown here. (Fuel selection and firing equipment information can be found in Chapters 14 through 18.

COMBUSTION CONTROLS, SAFETY CONTROLS AND INSTRUMENTATION

Combustion Controls

Combustion controls for HTW generators are similar to those used for steam generators, with the exception that discharge and inlet water temperatures are sensed and measured in lieu of steam pressure.

Since the HTW generators are

HOW TO SIZE HTW GENERATORS

In a typical application of HTW generators to the large shopping center mentioned, HTW water and chilled water are produced in a central plant and distributed in underground steel conduits to the various structures. Pipelines are housed in the basements of the buildings. See Figs. 2 and 3.

The given load factors are as follows:

Season or Time	Btu Required
Winter (heating, humidification and domestic hot water)	100,000,000
Summer (cooling, reheat and domestic hot water)	110,000,000
Between-seasons	50,000,000
Nights	30,000,000

The optimum HTW generator sizes to meet these conditions would consist of three generators: two 65,000,000 Btuh each; and one 45,000,000 Btuh unit.

This arangement would permit any two generators to handle the peak load at any period and the smaller unit would be adequate for the night loads. One of the two large units could handle the between-seasons load.

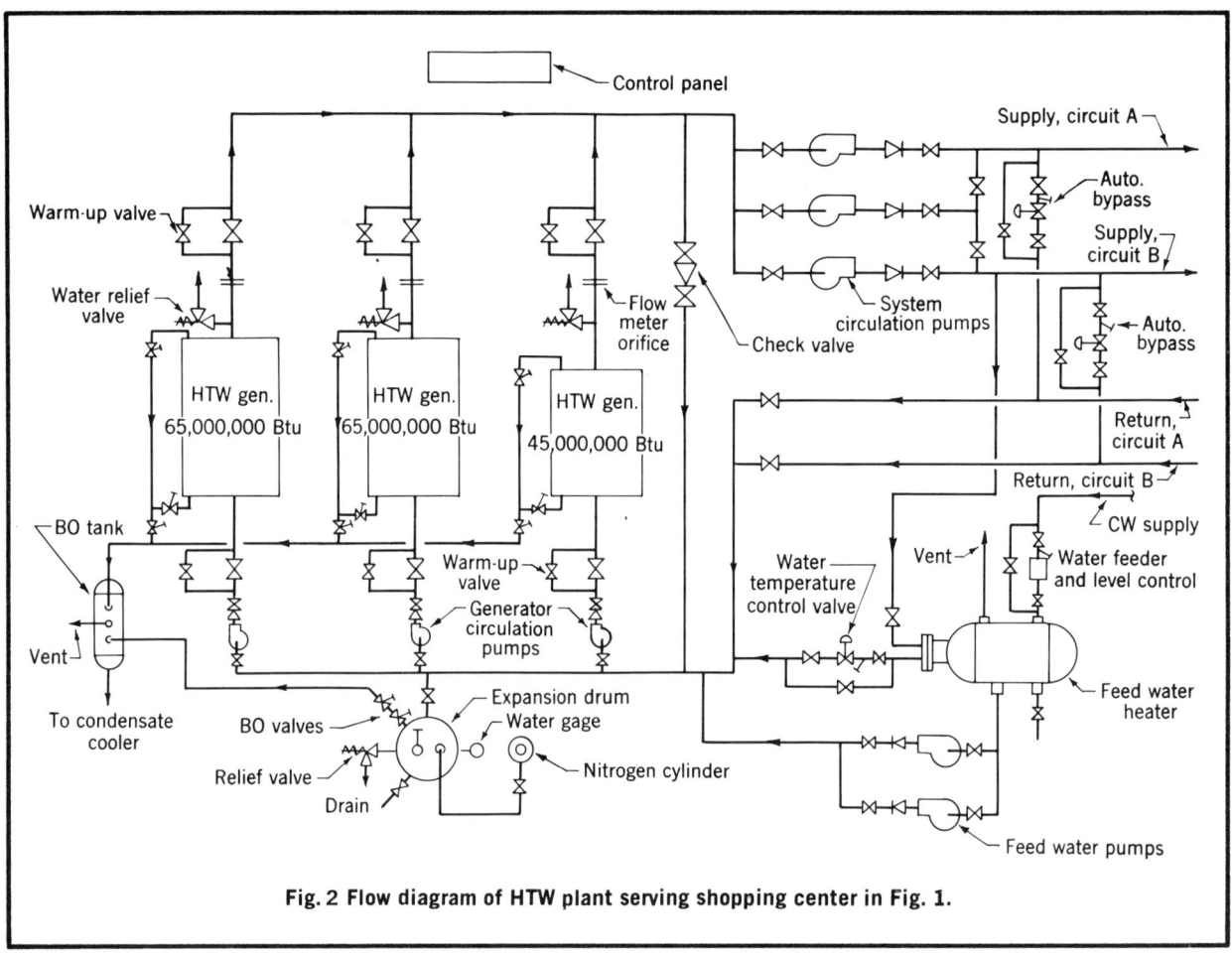

Fig. 2 Flow diagram of HTW plant serving shopping center in Fig. 1.

well above 20,000,000 Btu per hr in size, the combustion controls should be a full metering system.

The essential controls are: a master temperature controller; an air flow controller; and a fuel-air ratio controller.

Since discharge and inlet water temperatures tend to fluctuate, the master temperature controller compensates by sending a corrected signal to the cascade set point input of the air flow controller.

The air flow controller compares this loading signal with the corresponding metered air flow and sends corrective signals to the power unit to maintain the *correct air flow* loading at the inlet vanes of the forced draft fan.

The fuel-air ratio controller also uses air flow signals from

the air flow controller to govern the fuel valve opening or stoker speed, thus providing the *proper quantity of fuel*.

Safety Controls

Safety controls and safety control equipment must conform to Factory Insurance Association requirements to provide maximum safety.

HTW safety controls are also similar to those for steam generators, with the exception that water temperature and water flow enter the picture.

A control switch for low water flow is provided to stop fuel flow, if the generator water circulation rate falls below the preset safe point.

This control switch, usually furnished with an indicator, is the differential pressure type

and is connected across a flow orifice on the outlet line.

Instrumentation

Essential instrumentation should consist of: a Btu recorder to record heat gain of the HTW generator; pressure gages to indicate generator discharge and inlet HTW pressures; and temperature indicators to indicate discharge and inlet HTW temperatures at the generator.

Each HTW generator must also be equipped with an excess temperature cut-off.

Control Panels

Control panels should be free standing, canopy lighted and furnished with hinged doors at the rear. A single control panel may be provided for each generator, or an integrated master control panel may serve the total plant.

BOILER CIRCULATION AND SYSTEM PUMPS

An end suction, vertically split, cast steel pump lends itself ideally to the circulation of HTW for the generator and system.

The large, direct water entrance to the impeller materially reduces the possibility of cavitation because the impeller is evenly loaded with minimum turbulence. (The impeller is over-hung and attached to a large diameter shaft that rests upon two well spaced bearings.)

Mechanical shaft seals are not mandatory, but the author has found that they provide the simplest and most satisfactory method of shaft sealing. These mechanical shaft seals should be of the hydraulically balanced type, with stainless steel rotating parts. The hardened rotary seal face should be of tungsten carbide, and the mechanical seal chamber cooled and lubricated by a pumping ring in the seal, in conjunction with an external water-cooled heat exchanger.

(Packing will *not* stand up too well under high operating temperatures, and must be sealed and cooled by injection of filtered water which requires an additional pump, at a pressure higher than that of the system. Also, the added water would have to be blown down. All in all, a shaft packing is not too satisfactory.)

The pump should have pressure gages at both suction and discharge. Connections, bearings and seals should be protected by automatic flow control of cooling water.

Pump flanges must be protected from piping expansion strains by flexible stainless steel connections, inserted in suction and discharge piping at the pump.

The pump is driven through a spacer-type flexible coupling by an electric motor operating at 3450 or 1750 rpm.

How to Compute Pump Capacity

In the shopping center example previously cited, two circuits are used:

1. Circuit *A* with a load of 65,000,000 Btuh or 248,300 lb water per hr
2. Circuit *B* with a load of 45,000,000 Btuh or 171,900 lb water per hr.

Each circuit will have an individual automatic by-pass to keep each pump fully loaded under all conditions. Each HTW generator will have a circulating pump.

Circuit *A* pump capacity $= \dfrac{248,300}{51.55} = 4810$ cu ft water per hr

$$\frac{4810 \times 7.48}{60} = 600 \text{ gpm}$$

Circuit *B* pump capacity $= \dfrac{171,900}{51.55} = 3320$ cu ft water per hr

$$\frac{3320 \times 7.48}{60} = 414 \text{ gpm}$$

How to Compute Pump H.P.

Total friction loss (piping, valves, fittings, heat exchangers and control valves) is assumed to be 113 ft for circuit *A* and 161 ft for circuit *B*.

$$\text{Brake hp} = \frac{\text{gpm} \times \text{th} \times \text{sg of water}}{3960 \times \text{efficiency}}$$

where

gpm	= 600
th (total head)	= 113 ft
sg (specific gravity)	= 0.826 (450 F)
efficiency	= 80%

Pump *A* brake hp $= \dfrac{600 \times 113 \times 0.826}{3960 \times 0.8} = 17.68$

Actual pump selected is 20 hp.

Pump *B* brake hp $= \dfrac{414 \times 161 \times 0.826}{3960 \times 0.8} = 17.4$

Actual pump selected is 20 hp.

A 20 hp standby pump that could handle circuit *A* or *B* should be provided.

Variable Speed Drive

In many cases HTW systems for space heating area designed to permit expansion at a later date. Such an expansion calls for more water and higher pump heads. A pump can be selected to provide adequate discharge head for the original installation and for the anticipated expansion.

This larger capacity pump is connected to a variable speed drive and set to initial head requirements. When increased head is required, the pump speed is raised by adjusting the speed control—which may be of the magnetic or fluid type.

FEEDWATER HEATER

The HTW system is a sealed system, but some small water

losses occur at valve glands, flange areas and from possible failure of a mechanical seal. However small, these losses must be made up as needed to keep the system stable, and it should be done automatically.

Since cold make-up water must be heated to 211F to avoid thermal shock and to expel oxygen that must not be allowed to enter the system, the feedwater heater is an essential component . . . even though it is not in continuous use.

A simple feedwater heater that will do the job consists of a 180 gal horizontal tank equipped with a HTW heating coil, self-contained automatic sensing control to maintain water at 211F, thermometer, gage glass, feed make-up float control valve and a high-low level alarm. See Fig. 3.

The tank, fabricated of ⅜-in. thick steel plate welded construction, should sit upon a steel supporting frame at a height satisfactory for the make-up pump to handle 211F water without cavitation.

Because this tank is open to the atmosphere, it is not necessary that it be a code vessel.

Feedwater Make-up Pump

This pump must be of low capacity and high head in order to inject hot feed water into the HTW system against the total system pressure. This total system includes the following pressures:

1. saturation pressure of the water at its temperature

2. hydraulic head

3. circulating pump discharge pressure

4. nitrogen loading pressure

The total of these four pressures plus pump discharge piping friction plus a 5% safety factor, should provide adequate pump discharge head. (If the HTW system is steam pressurized, item 4 does not apply.)

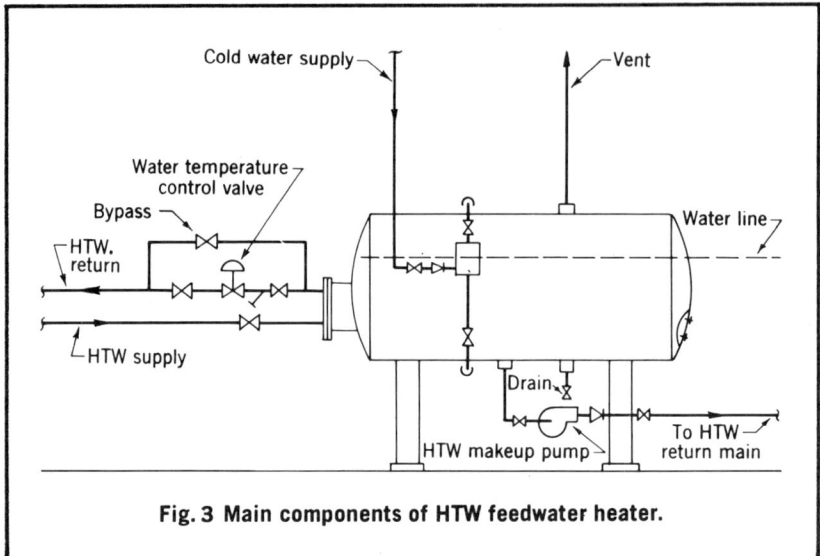

Fig. 3 Main components of HTW feedwater heater.

How to Calculate Feedwater Make-up Pump H.P.

In this example, the following values are given:

Water saturation pressure at 450 F	407.9 psig
System circulating pump discharge pressure	86.0
Nitrogen pressure	50.0
Make-up pump discharge piping pressure loss	10.0
	553.9 psig
Safety factor, 5%	27.0
Total pump discharge	580.9 psig

Make-up water requirement: 3 gpm
Specific gravity of water at 211 F: 0.9592

$$\text{Brake H.P.} = \frac{581 \times 2.31 \times 3 \times .9592}{3960 \times .80} = 1.211$$

Therefore, a 2 H.P. motor should be specified.

Feedwater Pump Control

The pump is automatically controlled by a water level controller (such as Magnetrol) set at the proper level at the expansion drum to provide cut-in and cut-out points.

As with the feedwater heater, some method must be adopted to prevent the feedwater make-up pump from injecting cold water into the system. One accepted way is an immersion aquastat and contactor with its sensor set in the feedwater tank and wired to the motor starter circuit. This arrangement prevents the pump motor from starting unless the feedwater temperature is 211F or higher.

MISCELLANEOUS CONSIDERATIONS

The water supply in most plants must be treated to remove excess mineral content.

Suppose that analysis of the raw water supply indicates a calcium hardness of 65 ppm of calcium carbonate ($CaCO_3$) and a magnesium hardness of 20 ppm of magnesium carbonate ($MgCO_3$) or a total of 85 ppm.

This type of hard water can

be softened by a sodium zeolite water treatment system (see water treatment in Chapter 13.)

The potential problem of carbon dioxide (CO_2) resulting from the bicarbonate during exchange of the calcium and magnesium carbonates in the HTW generator, and forming carbonic acid, is avoided by venting the carbon dioxide to atmosphere via manual vents.

HTW may be treated by using a by-pass feeder or chemical feed pump to inject sodium sulfite (to scavenge oxygen) and sodium hydroxide to reduce alkalinity.

If heat exchanger tubes are cupro-nickel, the pH factor must be maintained between 7 and 9.

If steel tubes are specified, a pH factor of 10 to 10.5 is recommended.

A water sample cooler should be provided.

Blow-off Tank and Condensate Cooler

A blow-off tank and condensate cooler are both necessary to contain and cool the blow-down. To size this equipment, refer to corresponding section on Steam Plants.

Plant Dimensions

The architect naturally looks to the mechanical designer to provide him with the necessary information as to length, width and height of the required building. The lay-out should allow adequate space for equipment, catwalks, platforms and ladders.

Clear headroom over the HTW generators and all other walk areas must be at least 7 ft. Preferably, the front and rear walls, at least, should contain as much window space as possible from floor to ceiling, with operable windows near floor level, and low silhouette gravity type ventilators on the roof, covering most of that area.

This arrangement will provide good, economical light and ventilation. Openings in end walls will provide space for gravity type combustion air heaters.

6

Expansion Drums

The HTW system must have some means to prevent damage to the piping or components due to expansion and contraction of the water.

One method to accomplish this is a steam pressurized expansion drum; a second method is a nitrogen pressurized expansion drum; a third approach uses a steam pressurized cascade heater which also functions as an expansion drum.

The cascade system, in conjunction with a steam generator is employed where a large amount of steam *and* HTW are required.

All of these vessels must conform to the pertinent ASME code for pressure vessels and be code stamped. The carbon steel plate used should be of fire-box quality, with an ultimate tensile strength of 70,000 psi and conform to ASTM specification SA-515-70. All nozzles must have steel flanges, and the drum must be all-welded construction for the required working pressure.

All vessels must carry insurance and inspection certificates from authorized pressure vessel inspection and insurance agencies.

Steam Pressurized Drum

In a steam pressurized system, the expansion drum is set hori-

How to Size a Steam Pressurized Expansion Drum

In this example, the following values are given:

System inlet temp: 200 F
System discharge temp: 450 F
System water content: 5500 gal
Water at 200 F: 60.132 lbs per cu ft
1 cu ft water = 7.48 gals
Water heated from 200 F to 450 F expands 0.00277 cu ft per lb
Since only half of system water will be at 450 F, expansion is based on 2750 gals.

$$\frac{2750}{7.48} = 368 \text{ cu ft water}$$

$368 \times 60.132 = 22,000$ lb water

Therefore, the expansion drum must provide additional space for expansion: $22,000 \times 0.00277 = 61$ cu ft

Assume drum diameter to be 6 ft with three 6-in. diameter water supply distributors along horizontal centerline.

Consequently, the top of the supply water distributor will be approximately 3½-in. above the centerline of the drum. Allowing 1 ft for expansion sets the waterline 15½-in. above the centerline. This leaves 20½-in. from the waterline to the top of the drum, which is adequate space. The 3-ft from the centerline to the bottom of the drum is space for the discharge water distributors.

Since the drum diameter is 6 ft, the rectangular volume for expansion will be 6 ft × 1 ft deep × drum length, an equivalent of 6 cu ft per lineal foot of drum.

Therefore, $\frac{61}{6} = 10$ ft 2½-in. minimum drum length.

To allow for small loss of space due to drum head curvature, the drum length should be 11 ft.

Steam safety valves on the drum provide complete blow-down and are set at a pressure 15% below that on the HTW generator. In addition, the drum is fitted with manhole, high pressure gage glass, thermometer, vent valve and blow-off valves.

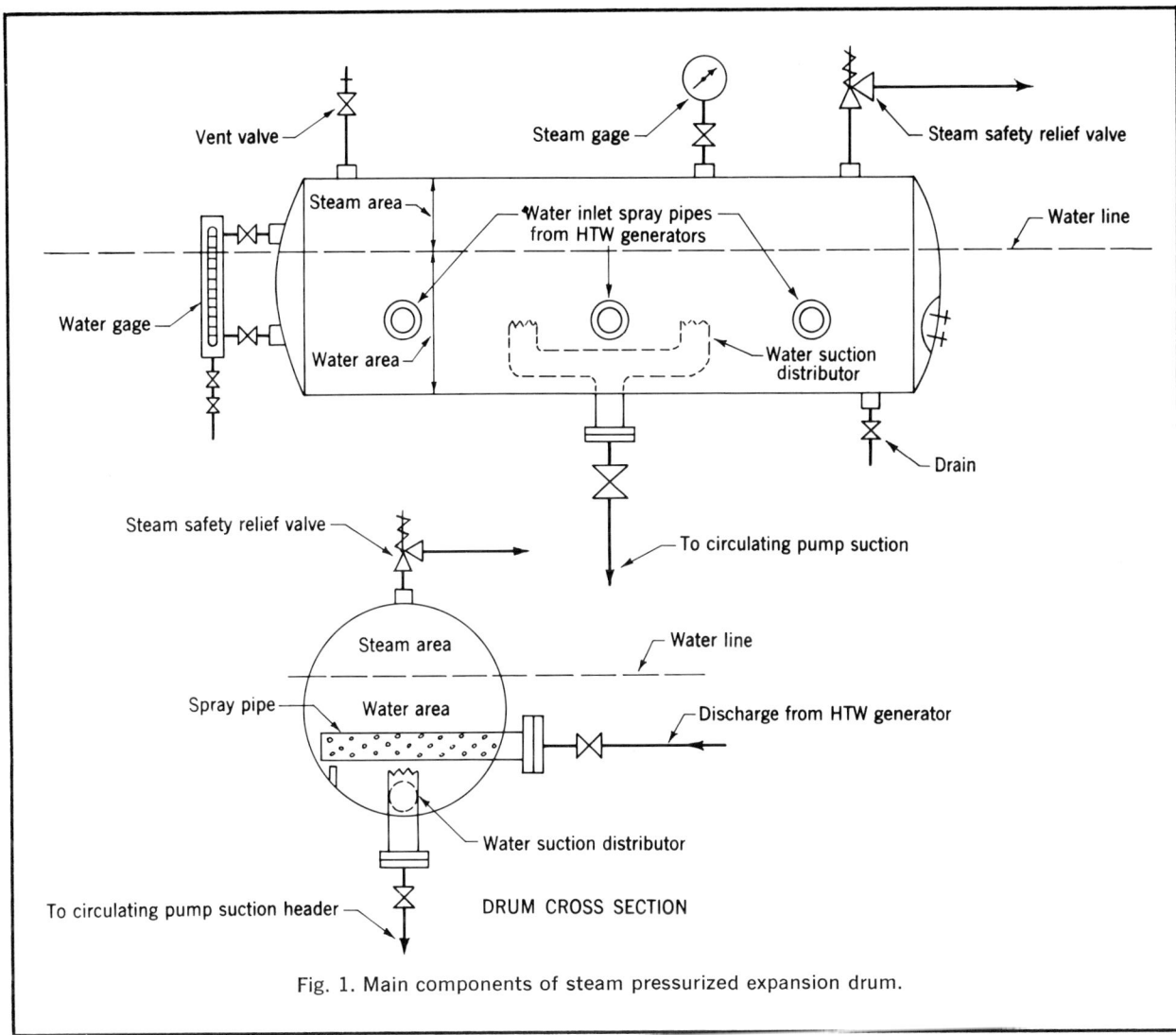

Fig. 1. Main components of steam pressurized expansion drum.

zontally and elevated well above the circulating pumps to prevent cavitation in the pumps. See Fig. 1.

The steam pressurized drum must provide adequate space above the center line for water distributors, for water expansion and for steam pressurization.

The drum is rigged much the same as the steam drum on a steam generator, with gage glass, controls, steam relief valves and a blow-off valve in the bottom, at one end of the drum.

Supply water distributors take up considerable space and must be properly sized. These distributors should have a diameter

How to Size a Nitrogen Pressurized Expansion Drum

In this example, the following values are given:

System inlet temp: 200 F (variable to 450 F)
System discharge temp: 450 F (constant)
System contents: 5500 gals
System design pressure: 500 psig (max)
Saturation pressure of water at 450 F: 408 psig
Water at 200 F: 60.132 lb per cu ft
Water: 7.48 gals per cu ft
Water heated from 200 F to 450 F expands 0.00277 cu ft per lb

Since only half of system water will be at 450 F, expansion is based on 2750 gals.

$$\frac{2750}{7.48} = 368 \text{ cu ft}$$

$$368 \times 60.132 = 22{,}000 \text{ lb water}$$

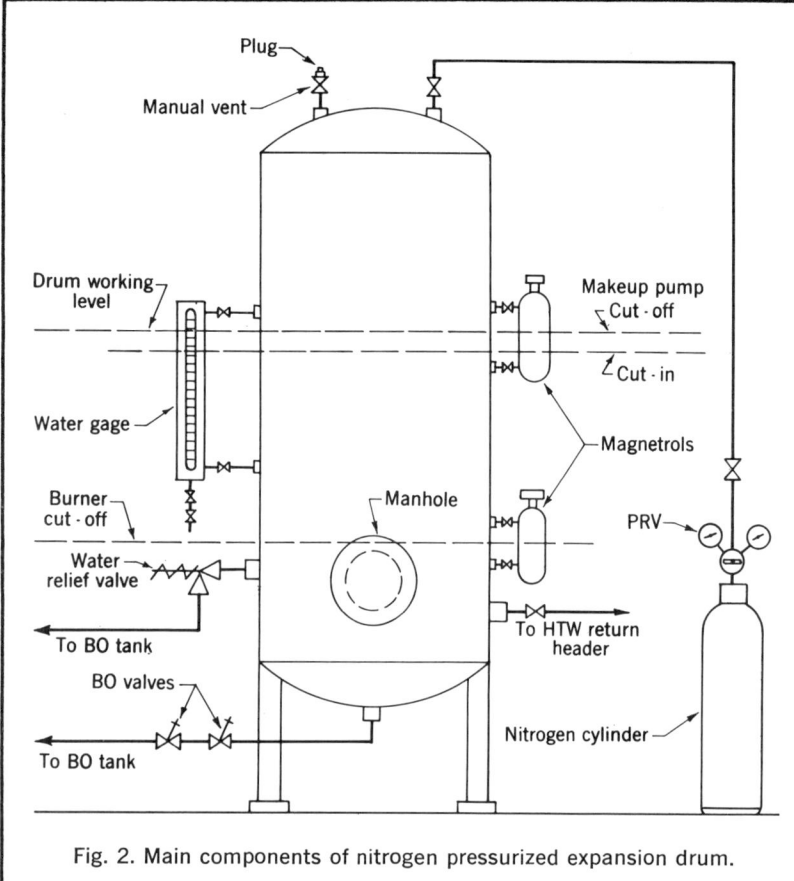

Fig. 2. Main components of nitrogen pressurized expansion drum.

large enough in diameter to provide a water velocity of not greater than 2 ft per second. Discharge distributors sit along the bottom centerline between the supply water distributors and project upward to a point below the drum working level.

Nitrogen Pressurized Expansion Drum

The nitrogen pressurized drum is the simplest of all the expansion arrangements described here. See Fig. 2.

It sits vertically on legs upon the operating floor, and is connected to the return side of the HTW system by a single pipe, which enters the side of the drum near the bottom.

Nitrogen is fed into the drum through a connection in the top head. Connections for controls, gage glass and water relief valve are on the side at the proper levels. A blow-off connection is taken from the center of the bottom head. The relief valve is set below the waterline for sealing, and to prevent the loss of nitrogen. However, a soft seated relief valve can be located at top of the drum.

Nitrogen Pressurization of Expansion Drum

There are several systems on the market today for nitrogen pressurizing the HTW expansion drum. These systems are expensive, complicated and unnecessary.

The HTW system can be pressurized with nitrogen simply by manually feeding the gas into the drum. All that is required is a cylinder of nitrogen fitted with a single stage pressure reducing valve, a delivery gage, and a cylinder inlet gage (a nitrogen cylinder contains 221 cu ft). See Fig. 2.

When a HTW system is put into operation, the drum is filled with water to the minimum operating level and nitrogen fed, manually, while the water in the system is heated to the design

Therefore, the expansion drum must provide space for:

    22,000 × 0.00277 = 61 cu ft
    61 × 7.48 = 457 gals expanded water

Assume drum to be 7 ft diameter × 18 ft high; then drum contains 38.4 cu ft or 288 gals per ft of height.

Therefore, expansion height $= \dfrac{457}{288} = 1.59$ ft

Allow 3 ft for connections, controls and minimum amount of water. Then height of nitrogen column = 18 − (3 + 1.59) = 13.41 ft.

Therefore nitrogen starting pressure $= \dfrac{15 \times (408 + 15)}{13.41} = 474$ psig

Drum safety valve should be set 10 lb higher at 484 psig.

equal to the generator outlet and have perforations equal to the cross-sectional area of the pipe. Supply distributors enter the steam pressurized drum on one side along the horizontal centerline and project across to the other side of the drum.

Discharge water distributors are open-ended and must be

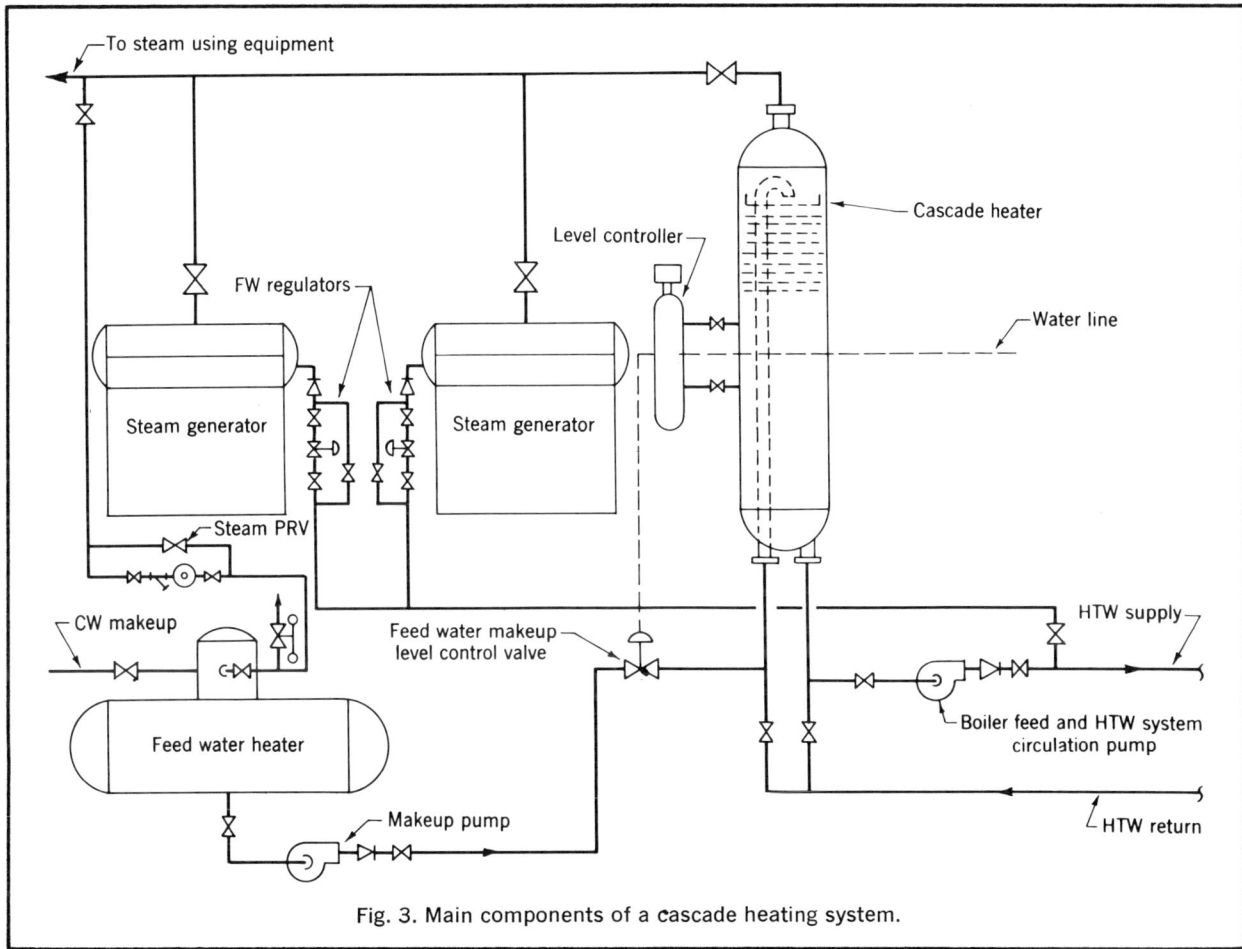

Fig. 3. Main components of a cascade heating system.

working temperature until the proper pressure level is reached. While this is occurring, the expanding water is blown down to maintain the proper working level in the drum. At equilibrium, the blow-off valve is closed.

From time to time, a small amount of nitrogen may have to be added to maintain the necessary pressure to keep the system above flashpoint.

Cascade Heater

The cascade heater is constructed vertically because it must contain water trays or sprays, or a combination of both.

It must be designed to provide adequate space for water retention, expansion, contraction, and steam volume, and is fitted out similarly to the steam loaded drum. See Fig. 3.

The system designer should consult with the cascade heater manufacturer to properly size this vessel.

AUTOMATIC HTW BY-PASS

The automatic HTW by-pass in the single zone system has two functions:

1. to constantly maintain adequate circulation through the generators as the heat users throttle down;

2. to maintain full flow through the circulating pumps to prevent cavitation in the pump chambers.

In the multi-zone system, where each HTW generator has its own circulating pump, the function of the automatic by-pass is to maintain adequate flow

through the zone pumps under all load conditions.

Components of the automatic by-pass system are as follows: D/P cell; orifice flange with stainless steel orifice plate; indicating receiving controller; air filter regulator; and double seated, proportional by-pass valve reactor.

The by-pass valve should be of industrial quality, and be pneumatically operated. The valve is spring loaded and, if air failure occurs, will come to full open position to protect the pumps, and/or HTW generators, and remain in full open position until air pressure is restored. The by-pass valve should be complete with strainer and blow-off valves and be set in a 3-valve manual by-pass arrangement.

7

High Temperature Water Heat Exchangers

Since HTW is the primary heating medium, its heat in most cases must be exchanged to a secondary water system to provide water or steam at lower temperatures and pressures.

This is done via a water-to-water heat exchanger, a simple device consisting of a steel shell, a steel tube sheet, U-shaped tubes, and a steel head and nozzles. See Fig. 1.

The tubes contain HTW and the shell holds secondary water. In the smaller shell diameters, steel pipe is used; in the larger diameters, steel plate is rolled and welded to form the shell. Heads are constructed of fabricated steel; tube sheet is made of thick steel plate that has been drilled, reamed and serrated to receive the tubes that are "rolled in".

The U-shaped tubes provide for expansion, and should be of steel or stress-relieved 90-10 cupro-nickel; certain water conditions and temperatures may require 70-30 cupro-nickel.

Heat exchangers should conform to ASME code requirements and carry the code stamp. The exchanger shell that carries the

secondary water is protected by a relief valve.

To produce steam, a coil is inserted longitudinally in the head of a steel drum near the bottom. A waterline is maintained at approximately drum centerline, leaving the space above for the accumulation of steam. Such an arrangement is known as an unfired steam generator. See Fig. 2.

The complete unit is skid-mounted and provided with:

HTW control valve to sense steam pressure; steam relief valves; water column with switch controls to operate feed pump and low water cut-off; main steam shut-off valve; blow-off valves; and a steam separator.

The following three examples show how to compute HTW Flow and Low Temperature Water (LTW) Flow for steam generators and heat exchangers.

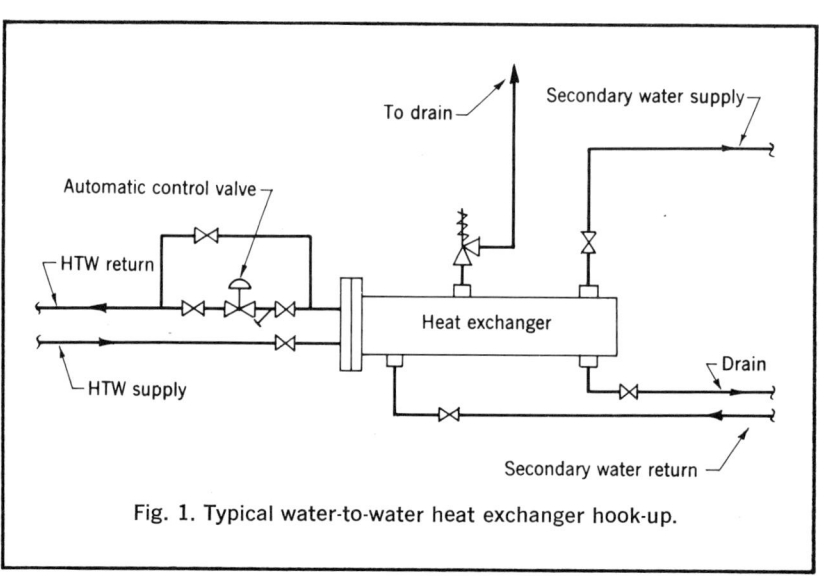

Fig. 1. Typical water-to-water heat exchanger hook-up.

HOW TO COMPUTE WATER FLOWS
HTW Flow for Steam Generators

In this example, the following values are given:

Quantity of steam to be supplied: 1000 lb/hr at 125 psig
Steam temp at 125 psig: 353 F
Steam condensate return temp: 150 F
Latent heat of steam at 125 psig: 868.2 Btu/lb
Sensible heat of steam at 125 psig: 325 Btu/lb
Sensible heat of steam condensate return at 150 F: 117 Btu/lb
HTW supply temp: 450 F
HTW return must be 10 F to 15 F above saturation steam temp
  (353 + 10) or 363 F

Average HTW temp: $\dfrac{450 + 363}{2} = 406.5$ F

HTW differential: $450 - 363 = 87$ F
HTW at 406.5 F is 7.15 lb per gal

Total heat input to heat exchanger is $Q_L + Q_S$ where

  Q_L = lb/hr steam × latent heat
  Q_S = lb/hr steam × difference between sensible heat of steam and
      condensate

Substituting given values in formula:

  $Q_L = 1000 \times 868.2 = 868,200$ Btu
  $Q_S = 1000 \times (325 - 117) = 208,000$ Btu
  Total heat input $= 1,076,200$ Btu

Specific heat of water at 450 F is 1.12
Specific heat of water at 363 F is 1.05

Mean specific heat is $\dfrac{1.12 + 1.05}{2} = 1.085$

$$\text{Water flow} = \frac{\text{Computed heat loss (Btuh)}}{\text{temp diff} \times \text{mean specific heat of water}}$$

substituting given values:

$$\text{HTW flow} = \frac{1,076,200}{87 \times 1.085} = 11,400 \text{ lb per hr}$$

Converting this figure to gpm $\dfrac{11,400}{7.15 \times 60} = 26.6$ gpm

HTW Flow for Water-to-Water Heat Exchanger

In this example, the following values are given:

Water supply temp: 450 F
Water return temp: 200 F
Average HTW temp: 325 F
Average LTW temp: 190 F
HTW differential: $450 - 200 = 250$ F
HTW at 325 F: 7.55 lb per gal
LTW at 190 F: 8.05 lb per gal
LTW flow: 519 gpm

$$\text{HTW flow} = \frac{\text{LTW flow} \times \text{temp diff} \times \text{wgt at avg temp} \times 60}{\text{HTW temp diff} \times \text{wgt at avg temp} \times 60}$$

Substituting given values:

$$\frac{519 \times 20 \times 8.05 \times 60}{250 \times 7.55 \times 60} = 44.5 \text{ gpm}$$

Specific heat of water at 450 F: 1.12
Specific heat of water at 200 F: 1.01

Mean specific heat is $\dfrac{1.12 + 1.01}{2} = 1.065$

Therefore, corrected HTW flow is $\dfrac{44.5}{1.065} = 42 \text{ gpm}$

LTW Flow for Water-to-Water Heat Exchangers

In this example, the following values are given:

Supply water temp: 200 F
Return water temp: 180 F
Average water temp: 190 F
Water at 190 F: 8.05 lb per gal
Computed heat loss: 5,000,000 Btuh

$$\text{Water flow} = \frac{\text{computed heat loss (Btuh)}}{\text{temp diff} \times \text{weight of water at avg temp} \times 60}$$

Substituting given values:

$$\text{LTW flow is } \frac{5,000,000}{20 \times 8.05 \times 60} = 519 \text{ gpm}$$

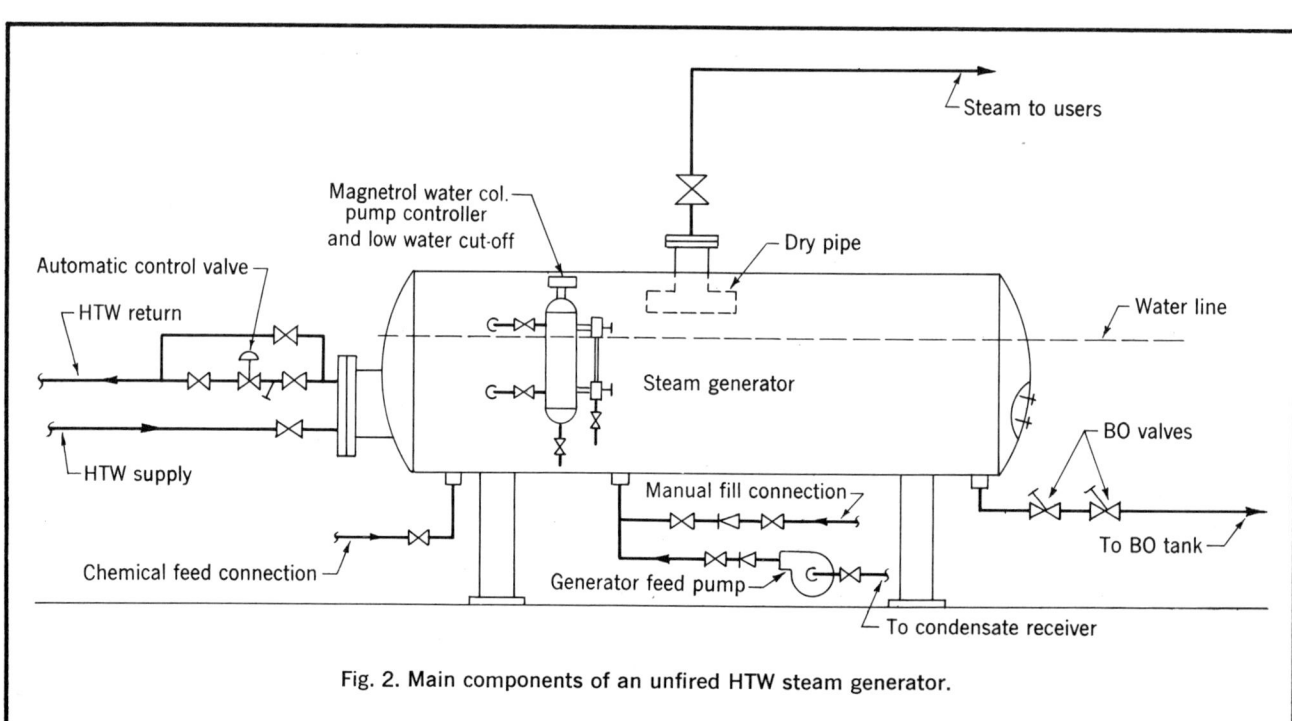

Fig. 2. Main components of an unfired HTW steam generator.

HEAT EXCHANGER CONTROL VALVES

Proportional control is the most satisfactory method of controlling the output of the heat exchanger. If temperature offset becomes too large to be acceptable, automatic reset should be added to readjust the control point to the set point.

Control valves should be pneumatically operated, normally closed, and of single seat design with tight close-off characteristics.

Heat exchanger control valves should be of industrial quality with flanged steel valve bodies, stainless steel trim, deep packing gland, top and bottom guided with equal percentage plug. Radiating fins may be desirable.

It is important to remember in the control of HTW used as a heating agent that the relationship between water flow and Btu's transferred through a heat exchanger is *non-linear*. Consequently, a flow reduction of 20% will reduce the heat transfer by 50%.

The controls must be designed as an integral part of the entire HTW system. To assure effective throttling performance by the control valves, the distribution system must provide uniform head to the control valves throughout the system and adequate pressure drop across all control valves.

HTW control valves may be installed in the return or supply side of water-to-water exchangers, and unfired steam generators.

Control valves should always be set in a three-valve, by-pass arrangement.

Proper sizing of the control valve is of the utmost importance, and to properly size the valve, three factors must be known:

1. design flow
2. inlet pressure
3. design pressure drop

When sizing control valves it is recommended that the valve manufacturer be called to assist because only they are completely familiar with all the characteristics of the particular valve.

8

Steam Distribution

A well designed, efficient piping system calls for proper materials, proper sizing, suitable insulation and installation that takes into account adequate support and allowance for expansion. (However, even with a good set of drawings, a satisfactory job will also depend on the skill of the pipefitters and the integrity of the job supervisors.)

Insulation, support and expansion, in particular, are related to whether the high pressure steam piping is installed in buildings, outdoors at ground level, below ground, or elevated.

For example, when piping is installed in buildings, it is suspended from the building construction. Outdoors, piping is set upon steel or concrete supports close to ground level or suspended from steel frames at some distance above ground.

Below-ground piping is run in walk tunnels, concrete box, steel or tile conduits; in insulating concrete conduit or cement asbestos conduit; or buried in dry,
water-impervious insulating material, wrapped in cellular glass insulation, etc.

Tunnels

When connecting 'pedestrian tunnels' (used in hospitals, universities and industrial complexes) are also used to carry distribution piping for steam and other utilities, the arrangement provides the most economical method of steam distribution and also provides easy access for maintenance and repair.

Single purpose 'walk tunnels' provide only sufficient space for pipe installation and maintenance and consequently pose a higher distribution cost than the multi-use pedestrian tunnel. The walk tunnel may be constructed in wet or dry soil.

Concrete Box Conduit

The concrete box conduit costs less than the walk type tunnel. Bottom and side walls are cast in place and the top is constructed of precast concrete slab
sections. Ship-lap joints and eye hooks cast in each section permit easy removal of the top sections. Roof slab sections should not be longer than 3 ft. Piping should be covered with sectional insulation and supported on rolls that rest upon chair carriers.

If expansion loops are used, the conduit must be extended to encompass them. Access pits must be supplied for corrugated, slip or ball type expansion joints and for valves, drip traps and condensate pumps.

Conduit construction should be 2500 psi concrete with a minimum thickness of 4 in. This conduit should be pitched to a drain point and insulated with calcium silicate. Top and sidewalls should be covered with hot pitch and layers of waterproof membrane.

This type of conduit should not be used where the water table is above the conduit base.

Underground Steel Conduit

Underground steel conduit construction of heavy gage steel

is a popular, medium-cost method for protecting underground pipe line.

To provide a stronger conduit, some manufacturers use spiral corrugations to stiffen the steel wall. Other variations are spiral welded steel strips, smooth-walled with internal stiffening, section connections with flanges, and welded connnections.

All steel conduits should be heavily zinc-coated inside and outside, using a minimum of 2 oz zinc per sq ft.

Outside surfaces should be given a thick coat of hot bitumastic compound and felt. Some manufacturers add a layer of fiberglass cloth.

Pipe support usually consists of interior vertical steel plates set at right angles to the conduit run.

Piping is insulated in the conventional manner except that the inside periphery is sometimes lined with insulation for special uses. A 1-inch annular airspace must be provided between conduit wall and the pipe insulation.

Expansion of pipe may be provided by pipe loops or mechanical joints — ball, slip or corrugated type. If loops are used, the conduit is oversized in the loop area to accommodate the pipe movements. Mechanical joints require an access chamber. Space for lateral pipe expansion (where the direction changes) is provided by enlarging the conduit in that area.

Piping is anchored by steel plates set at right angles to the conduit and welded to the pipes and conduit so that they extend beyond the conduit. A concrete anchor is then poured completely around the conduit and plate.

All steel conduits should have airtight end seal plates and be air tested to at least 15 psig.

The engineer should assume that all conduits will leak at some time; therefore, conduits must be pitched to a drain point,

and the piping insulated with calcium silicate, which, when wetted, will not break up, and may be dried out without damage.

These steel conduits may be installed in wet or dry soil. For passage under railroads, heavy traffic arteries or streams, the conduit can be supplied in cast iron.

Sacrificial zinc anodes must be installed on all-steel conduits to provide maximum protection against electrolytic action.

Insulating Concrete Conduit

This type of conduit is completely field fabricated. Constructed of vermiculite, cement and water, the top, sides and bottom are wrapped with sheets of polyvinyl chloride to a minimum thickness of 20 mils.

Insulated concrete conduit will handle pipe temperatures to 1200F and works well in dry soils. The load carrying capacity is 10,000 lb per linear ft (approx.), which provides a high safety factor.

Concrete support blocks are set upon a 4-in. thick reinforced concrete base pad. Before laying the pipe on these blocks, it is wrapped in a single layer of corrugated paper to prevent adherence to the vermiculite concrete. After the pipe is laid, it is welded and hydrostatically tested at 1.5 times operating pressure.

After passing the hydrostatic test, vertical forms are installed along the sides of the base pad and the exterior polyvinyl chloride envelope is placed inside the forms and under the concrete support blocks. Insulating concrete is then poured and, after setting, the PVC envelope is closed to completely cover the conduit.

Just above the base pad, near the outer periphery of the conduit, drain holes are formed throughout the length of the conduit by setting plastic, expendable tubes filled with water be-

fore pouring the concrete. After the concrete has set, the water is drained off, the tubes are collapsed and removed, leaving a continuous row of drain holes.

If pipe loops are used for pipeline expansion, the required voids in the concrete are created by inserting sheet metal enclosures before pouring the concrete.

If mechanical expansion joints are used in an underground system, concrete chambers must be constructed to house them and to provide access.

To anchor the pipeline, the base pad is thickened and heavy reinforcing bar is installed. Then steel plate is installed with the pipe run and is welded to the reinforcing bar and pipe to form the anchor. Anchoring at pipe elbows is accomplished in the same way.

Insulating Hydrocarbon Conduits

There are two materials used to make hydrocarbon type insulating conduits:

1. Gilso-Therm 70*, known as Gilsonite, a dry crystalline material
2. Asphalt, an air-blown petrochemical product. Both are recommended for temperatures ranging from 35F to 470F.

The k factor (Btu/hr/sq ft/deg F/in) for the following mean temperatures is:

Mean temperature 50 F, $k = 0.54$

Mean temperature 175 F, $k = 0.66$

Mean temperature 260 F, $k = 0.85$

Gilso-Therm 70 is installed dry without sintering. First, a trench is excavated and the pipeline set upon steel, transite or concrete supports. Anchors are con-

---

*Gilso-Therm 70 is the successor to Gilso-Therm, Gilso-Gard and Gilsulate, all products of The Gilsonite Company.

structed of structural steel or steel plate and enclosed in a concrete anchor block. Then the Gilso-Therm 70 is poured into the trench, under and over the pipeline to the required thickness, after which the trench is backfilled. This type of conduit is recommended for dry soils only.

The petroleum type asphalt is mixed with Pearlite to form the insulating material and is installed in the same way as insulating concrete conduit.

Cellular Glass Conduit

This inorganic, rigid material is suitable for temperatures to 350F and has a compression strength of 100 psi. Cellular glass conduit is recommended for areas where ground water is not a problem.

Supplied in premolded half-sections, cellular glass conduit is applied to the pipe and waterproofed by placing alternate layers of cut-back asphalt coating (3 layers) and asphalt-impregnated glass fabric (2 layers).

The insulated pipe is placed in a trench upon a 3-in sandfill. If there is some indication of ground water, drain tile or a stone leaching fill should be placed in the trench before putting in the sand bed.

Fill, above the conduit, should be 3 in. of sand and at least 12 in. of selected material free of large stones. The remainder of the fill can be the usual site material.

Cement-Asbestos Conduit

Cement-asbestos conduit is very similar to steel conduit. Constructed of a mixture of cement, asbestos and water, the conduit is formed under high hydraulic pressure. Its crushing strength ranges from 2400 to 3300 lb per lin. ft, depending on diameter.

Conduit sections are connected by split cement-asbestos couplings and made tight with a bonding material. After installation, conduit is air tested at 15 psi.

Space for expansion loops is provided by using oversize pipe covering, set upon a concrete base and covered with 3-in. thick concrete. Pipe anchors are constructed of conventional structural steel or steel plate, and are encased in concrete anchor block.

Clay Tile Conduit

There are two types of clay tile conduit: full round tile laid without base in the trench; and arch tile built upon a concrete base.

Full round clay tile is supplied in round 2-ft sections with bell and spigot, and scored each side along the tile at its mid-section. It is split at the site, laid in the trench and the piping installed on supports in the half section.

Cement mortar is used to grout the sections together. Elbows are used for changes in direction and are oversized for expansion loops.

This type conduit is pitched and the lower half is used as a drain. Sometimes, a base tile with drainage slots is furnished. Preformed and loose fill insulation may be used, but preformed calcium silicate insulation is preferred. Full round clay conduit should be used only in dry soils.

Arch or half round tile is composed of three sections, two side blocks and an arched top. The side blocks are laid upon a concrete base, in which a drain channel runs down the center. The side blocks are grouted to the base, and to each other. Pipe supports are set upon the concrete base, and the arch tile grouted into place. Preformed calcium silicate is also preferred as insulation. Conduit sections are enlarged for expansion loops and sections are mitered for change of direction.

Joints are sometimes coated with asphalt and a waterproof membrane.

Anchoring is provided by using structural steel members cast into the concrete base (thickened at the anchor point) and welded to the pipe line.

Guiding, where required, is handled in the same manner. The conduit should be pitched to the drain point.

Arch tile conduit provides a good housing for pipelines laid in dry soils.

Conduit Piping

All high-pressure steam piping should be welded black carbon steel construction, never less than Schedule 40 and heavier wall thickness when required. Bent or coiled piping should conform to ASTM specifications, designation A53, grade A.

Condensate return piping may be Schedule 80 steel or wrought iron. Copper or brass pipe should not be employed because of difficulties in obtaining electrical isolation.

Welded fittings should be of the same schedule as the particular pipe employed. All welding should conform to American Welding Society standards.

Flanges should be steel, weld neck or slip-on types. Slip-on type must be welded inside and outside.

Valves

Steam valves used at pressures up to 250 psig may be bronze or cast iron flange or screw type, as required. Above 250 psig, valve construction should be steel and be flanged or welded, as required.

Pipe Expansion

Expansion of steam and condensate lines, whether above ground, underground or in buildings, is controlled by loops or by slip, corrugated and ball type expansion joints. The two main factors that influence the

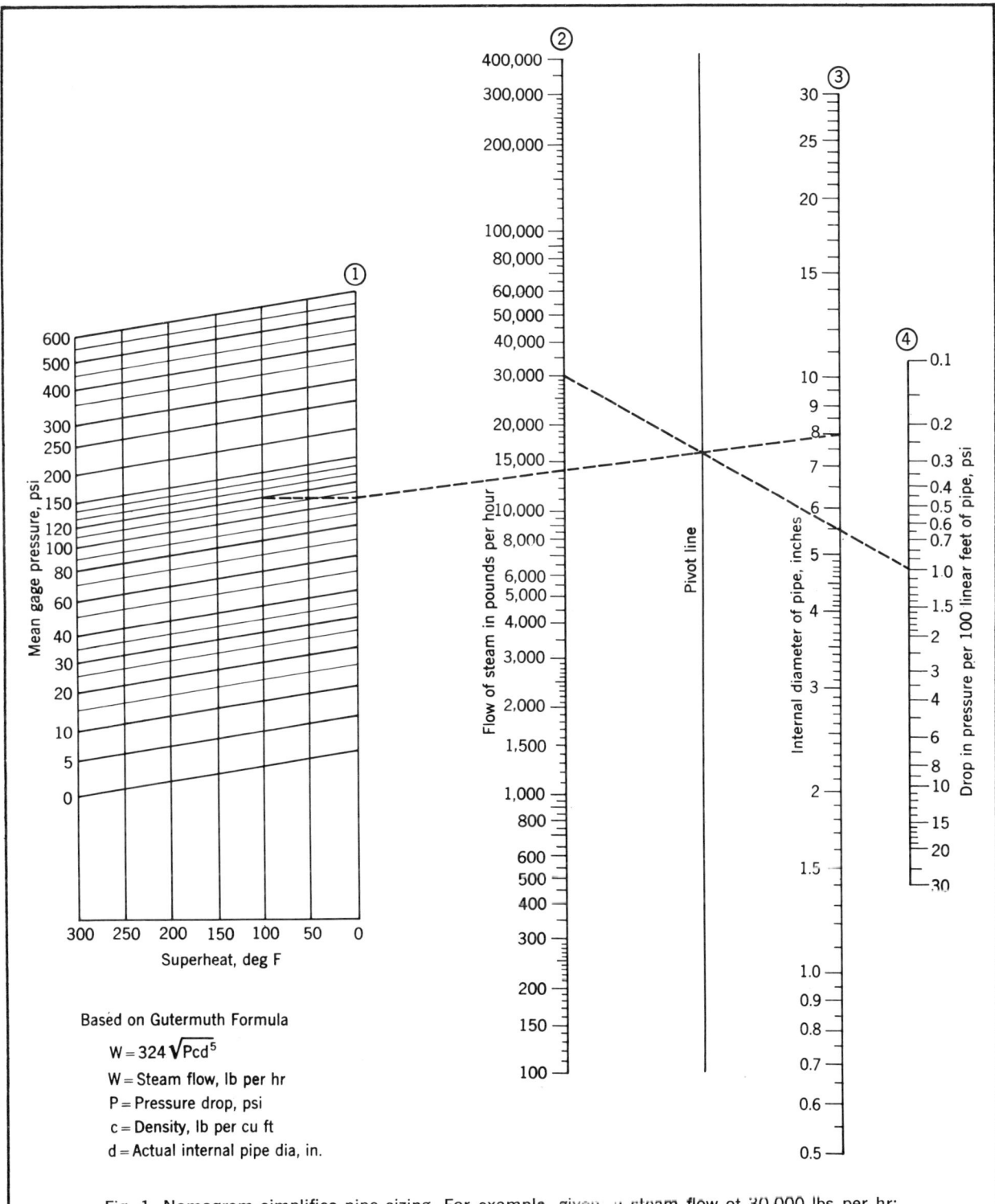

Based on Gutermuth Formula

$$W = 324 \sqrt{Pcd^5}$$

W = Steam flow, lb per hr
P = Pressure drop, psi
c = Density, lb per cu ft
d = Actual internal pipe dia, in.

Fig. 1. Nomogram simplifies pipe sizing. For example, given a steam flow of 30,000 lbs per hr; initial pressure 125 psig; superheat 100 F, final pressure desired 115 psig; equivalent length of pipe 1000 ft; pressure drop 1 lb per 100 ft; average pressure 120 psig. To find necessary pipe diameter: join 30,000 on Scale 2 to 1 on Scale 4. Locate intersection of 120 lb diagonal and 100 F vertical and proceed horizontally to Scale 1. Project line from this point on Scale 1 through intersection on Pivot Line to Scale 3. Answer from Scale 3 is 8-in diameter pipe is required. Courtesy of and copyright by McGraw-Hill, Inc. Reprinted with permission from POWER, December 1933.

designer's choice are economics and space. Generally, the choice is between the pipe expansion loop and mechanical joint.

If a mechanical joint is considered for underground distribution, where no valves are employed, the cost of an access chamber must be taken into account.

Pipe Sizing

Proper sizing of piping for a high pressure steam plant is very important, both from an economic and engineering viewpoint.

Obviously, piping that is larger than necessary is wasteful but, in piping that is too small, the pressure drop is so great that steam pressure at the equipment, or working end, falls below the required figure and return condensate will be insufficient.

Two guides will be found useful in determining the correct pipe sizing that will yield acceptable steam pressure drop and steam velocity are the Nomogram (Fig. 1) whose use is explained by the accompanying caption, and the formula for velocity of steam flow in pipes.

The velocity formula is:

$$V = \frac{A \times B \times 1728}{C \times 60 \times 12}$$

$$= \frac{A \times B \times 2.4}{C}$$

where

V = velocity of steam, fpm
A = lb of steam per hr
B = volume of steam at given gage pressure, cu ft per lb (see Steam Tables)

C = area of pipe, sq in.
1728 = cu in. per cu ft
60 = minutes per hr
12 = in. per ft

Therefore

$$A = \frac{C \times V}{B \times 2.4}$$

$$B = \frac{C \times V}{A \times 2.4}$$

$$C = \frac{A \times B \times 2.4}{V}$$

Application of the nomogram and formula is demonstrated in the following example:

Steam flow = 200,000 lb dry saturated steam per hr

Initial pressure = 300 psig

Final pressure required = 250 psig

Average pressure required = 275 psig

Length of pipe = 500 ft

Specific volume of steam at 300 psig = 1.4723

Applying these values to the pipe sizing nomogram shows that a 10-in Schedule 40 pipe would result in a pressure drop of 5 lb per 100 ft, or a total pressure drop of 25 lb for the 500 ft pipe and, therefore, a final pressure of 275 psig, which is quite adequate.

Using the velocity formula:

$$V = \frac{A \times B \times 2.4}{C}$$

$$= \frac{200,000 \times 1.4723 \times 2.4}{79}$$

$$= 9000 \text{ fpm}$$

Since acceptable velocities of saturated steam at pressures of 50 psig and higher can be 6000 to 10,000 fpm, the value of 9000 fpm is reasonable.

Suppose an 8-in. Schedule 40 pipe were considered; then, the resulting pressure drop would be 17 psig per 100 ft or 85 psig for the 500 ft line. This would result in a final pressure of 300-85, or 215, psig, which is 60 psig less than the required pressure.

The velocity for the 8-in. pipe, 14,200 fpm, is not acceptable and the corresponding pressure drop is excessive.

Short steam lines, such as headers, where the pressure drop is not significant, can be sized on a velocity basis. In such cases, the velocity may easily run to 10,000 fpm or even higher.

Steam generator lead velocities should be in the range of 7000 to 9000 fpm. This range will not cause an excessive drop in pressure and at the same time will prevent chattering of the disc in the stop check valve.

Superheated steam velocities can be 7000 to 20,000 fpm.

In general, the lower velocities should be used for pipes 12-in. dia and less. Higher velocities are used in pipes larger than 12-in. in dia.

Condensate return to the steam plant (return pump rate) should be at least twice as fast as steam flow from the plant.

Pipe wall thickness for pumped condensate lines should be at least Schedule 80 or larger.

Sizing the condensate return lines may be accomplished by using Cameron Hydraulic Data Tables. In such cases, water velocity and pressure drop must be carefully taken into account.

See Chapter 10 for further piping details.

High Temperature Water Distribution

DISTRIBUTION OF HTW

One of the many advantages of HTW piping is that it may be run perfectly flat without pitch; uphill and downhill, following the natural grade; or in buildings where construction requirements cause lines to rise or drop.

The only requirement is that an air bottle, with vent valve, be placed at the high points to manually vent the pipe line. See Fig. 1.

The distribution of HTW piping is handled in the same manner as high pressure steam piping, with the exception that the weight of the water must be considered (refer to Chapter 8).

HTW Piping

Pipe, for use with HTW, should never be less than Schedule 40 thickness, and may be thicker, as required.

Mild black carbon steel pipe should be used, and where pipe is bent or coiled, it should conform to ASTM specifications—Designation *A* 53, Grade *A*.

All pipe should be welded in accordance with the requirements of the American Welding Society and ASME Code. *Do not use threaded pipe.*

Welded fittings should have the same wall thickness as the pipe and be of the same material.

Flanges should be steel and may be of the welded neck or slip-on types. If slip-on type, they must be *welded inside and outside.*

Piping used in HTW plants

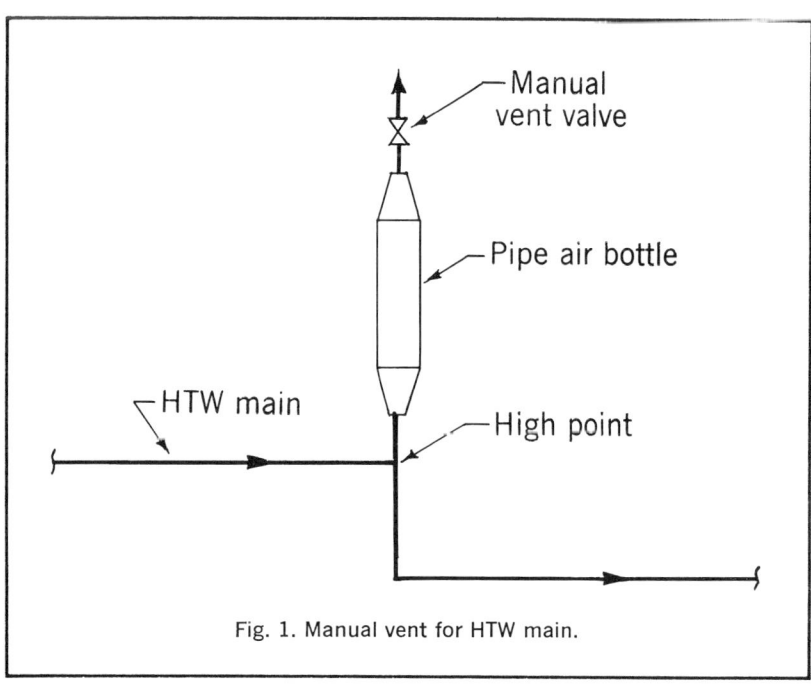

Fig. 1. Manual vent for HTW main.

45

should hang from spring hangers and, where vibration is a factor, vibration isolators should also be installed. Spring hangers should also be used on piping runs where expansion causes pipe movement. (For details on pipe hangers, see Chapter 10.)

Valves

Valves used for HTW must be constructed of cast steel and have a stainless steel seating and trim.

Steel valves 2-in. and larger may be flanged or welded.

Steel valves 1½-in. and smaller must be socket welded.

Use only gate or globe type valves; *screwed type valves must not be used* because they are a potential source of leaks.

When consulting a valve catalog, the designer should note that the working pressure at the rated top temperature for a HTW valve is much higher than for the corresponding steam type valve.

Pipe Expansion

Expansion of HTW lines due to temperature changes is handled by expansion loops.

Although a ball type expansion joint may be employed in some cases, the use of corrugated and slip type expansion joints pose problems.

For example, a slip type joint can be repacked under operating pressure, but other repairs to this type joint require the system to be shut down.

The corrugated type expansion joint, due to temperature changes, is subject to fatigue after a limited number of cycles.

For sizing expansion loops, see Chapter 10.

PIPE SIZING

One of the reasons for economy of a HTW plant is the use of smaller size pipe lines than those used in a high pressure steam plant supplying the same Btu quantity. And it is up to the designer to specify the most economical pipe size.

There is a tendency, at first,

Table 1. Influence of Temperature Differentials on selection of pump and Pipe Sizes for HTW Systems

	Temperature Differential				
	20	50	100	150	200
Generator discharge temp, F	270	300	350	400	450
Generator return temp, F	250	250	250	250	250
Flow rate for 20,000,000 Btuh load, lb/hr	1,000,000	400,000	200,000	133,000	100,000
Density of returning water, lb/gal	7.86	7.86	7.86	7.86	7.86
Pump capacity, gpm	2125	850	425	283	212.5
Assumed pump head, ft	100	100	100	100	100
Pump HP required, hp	84.0	33.6	16.8	11.2	8.4
Pump efficiency, %	60	60	60	60	60
Distribution pipe size, in.	8	6	4	3½	3

for the designer to over-size the piping, not realizing that under higher temperatures and pressures, water is lighter and less viscous than the low temperature hot water to which he has been accustomed.

What must be realized is that the quantity of Btu's per pound of water varies according to the design temperature differential.

For example, a system operating under full load at 450F with a return water temperature of 200F will send out 262 Btu per pound water. Operating at 400F and 250F respectively, each pound of water sends out only 158.5 Btu. See Table 1.

The reason for this inconsistency is that the *weight* of a cu ft of water varies inversely with its temperature. At 450F, a cu ft of water weighs 51.55 lb, but at 250F the water weighs 58.824 lb.

However, the *volume* of water per cu ft is constant. Therefore, the number of Btu's per *gallon* of water will vary relatively and absolutely according to its temperature.

Consequently, pipeline sizes are based on pounds of water per hour rather than gallons per minute. This practice is also due to the difference between the U.S. and Imperial gallon.

The following example shows how to convert lbs of water per hr to gals per min.

How to Convert Water Flow in lbs per hr to U.S. gals per min

In this example, the following values are given:

Total plant load: 100,000,000 Btuh
Supply water temp: 450 F
Return water temp: 200 F
Temp differential: 250 F
Heat yielded by water: 262 Btu per lb
Water at 450 F: 51.55 lb per cu ft
Water: 7.48 gal per cu ft
Water flow is

$$\frac{100,000,000}{262} = 382,000 \text{ lb per hr} = \frac{382,000}{51.55} = 7400 \text{ cu ft}$$

$$\frac{7400 \times 7.48}{60} = 922 \text{ gal per min}$$

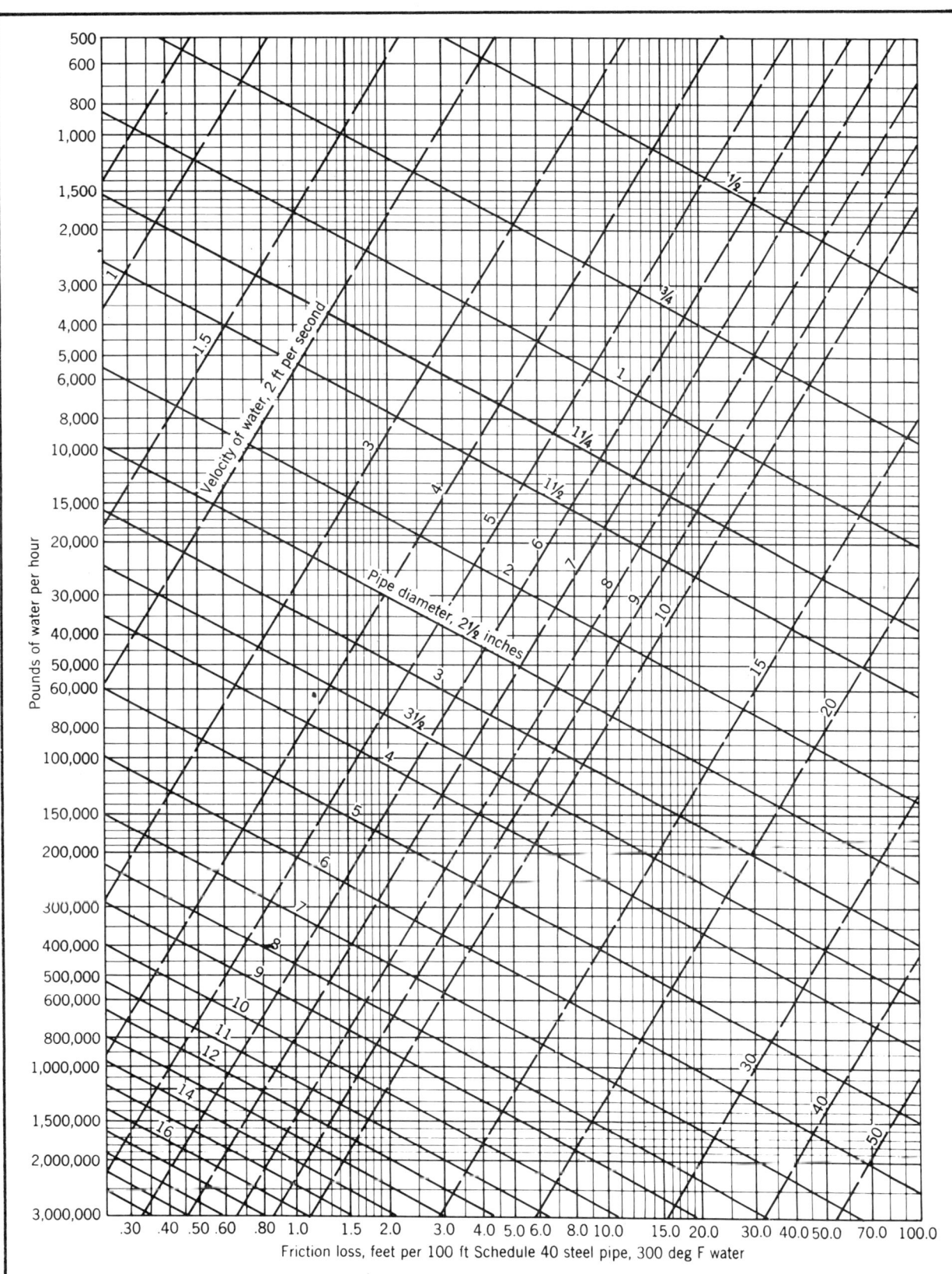

Fig. 2. Pounds of water per hour versus friction loss, feet per 100 ft Schedule 40 steel pipe, 300 deg F water.

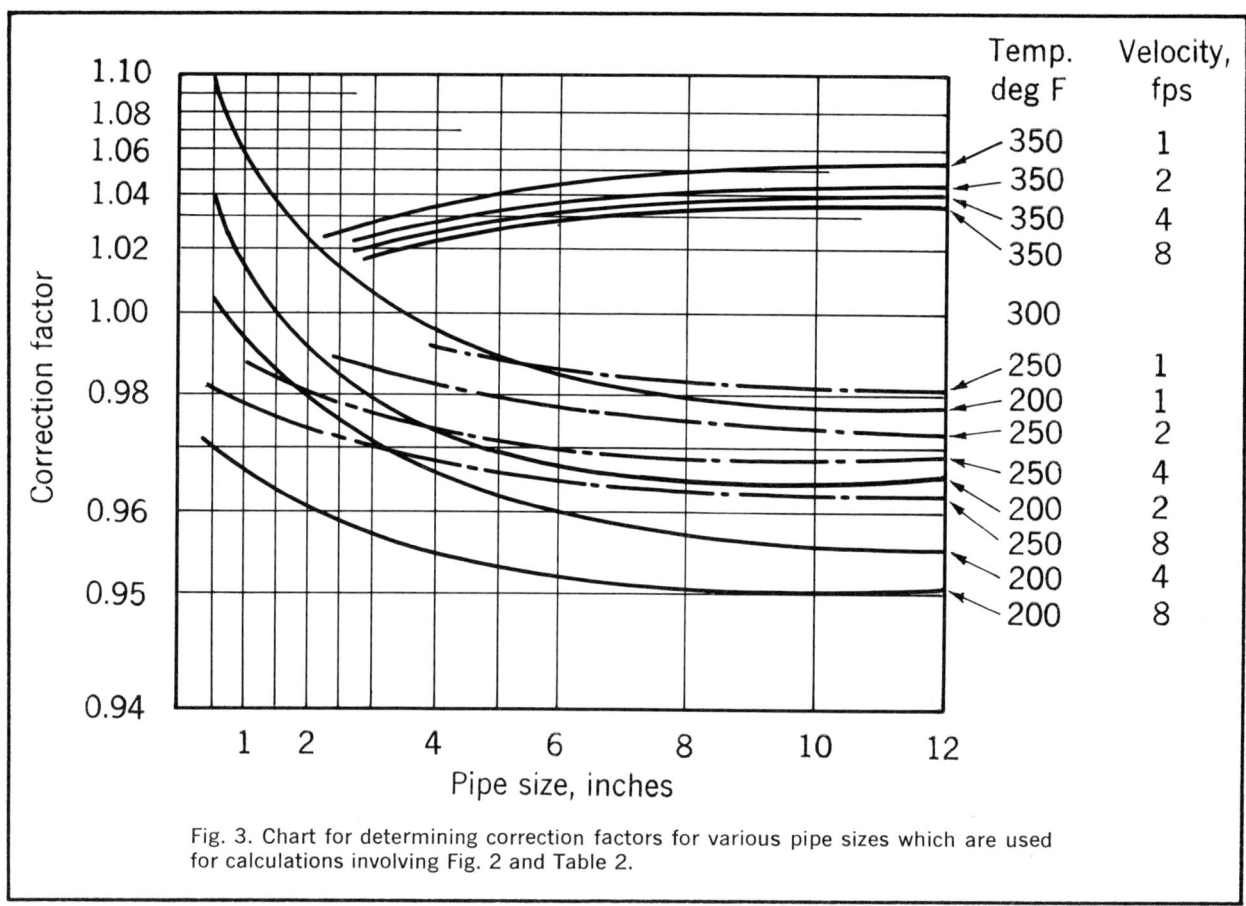

Fig. 3. Chart for determining correction factors for various pipe sizes which are used for calculations involving Fig. 2 and Table 2.

Table 2. Frictional Equivalent Length of Pipe Fittings
(Schedule 40 Steel Pipe, Turbulent Flow)

	Nominal Pipe Diameter, Inches													
	½	¾	1	1¼	1½	2	2½	3	3½	4	5	6	8	10
Fitting	Equivalent Length, Feet of Straight Pipe													
Globe Valve, Open	20.7	27.5	34.9	46.0	53.6	68.9	82.3	102.3	118.3	134.2	168.2	202.2	226.0	334.0
Angle Valve, Open	10.2	13.4	17.1	22.5	26.3	33.7	40.3	50.1	57.9	65.7	82.4	99.0	130.2	163.5
Swing Check Valve	5.0	6.6	8.5	11.1	13.0	16.7	19.9	24.7	28.6	32.4	40.7	48.9	64.3	80.7
180° Closed Screwed Return	4.3	5.7	7.2	9.5	11.1	14.2	17.0	21.1	24.4	27.7	34.7	41.7	54.9	68.9
Screwed Tee Through Branch	3.7	4.9	6.3	8.2	9.6	12.3	14.8	18.3	21.2	24.0	30.1	36.2	47.7	59.8
Welding Tee Through Branch	3.0	3.9	5.0	6.6	7.7	9.9	11.8	14.7	17.0	19.3	24.2	29.1	38.2	48.0
Submerged Discharge	2.2	2.9	3.7	4.9	5.8	7.4	8.8	11.0	12.7	14.4	18.0	21.7	28.5	35.8
90° Screwed Elbow	2.0	2.7	3.5	4.6	5.3	6.8	8.1	10.1	11.7	13.3	16.6	20.0	26.3	33.1
45° Lateral, Through Branch	1.6	2.1	2.7	3.6	4.2	5.4	6.4	8.0	9.2	10.5	13.1	15.8	20.8	26.1
90° Long Sweep	1.3	1.6	2.1	2.8	3.2	4.2	5.0	6.2	7.1	8.1	10.2	12.2	16.1	20.2
Screwed Tee Through run Lateral Tee Through run Submerged Entrance	1.1	1.4	1.8	2.4	2.8	3.4	4.3	5.3	6.2	7.0	8.8	10.5	13.9	17.4
180° Welding Return or bend R/d = 1.6 45° Screwed Elbow	.90	1.2	1.5	2.0	2.3	3.0	3.6	4.5	5.2	5.9	7.4	8.8	11.6	14.6
Welding Tee Through run	.82	1.1	1.4	1.8	2.1	2.7	3.3	4.0	4.7	5.3	6.7	8.0	10.5	13.2
90° Welding Elbow R/d 1.6	.68	.9	1.1	1.5	1.8	2.3	2.7	3.4	3.9	4.4	5.5	6.6	8.7	11.0
45° Welding Elbow Gate Valve, Open	.45	.6	.75	.99	1.2	1.5	1.8	2.2	2.6	2.9	3.6	4.4	5.7	7.2

How To Size Pipe

The factors that determine pipe size are: quantity of water flow, water velocity and frictional resistance of the pipe, valves and fittings.

On the pipe sizing chart, quantity of water is expressed as pounds of water per hour, velocity as feet per second and friction or head loss as feet per 100 feet. See Figs. 2, 3 and Table 2.

If the water velocity is excessive, the result will be high friction loss leading to high pump head and high pump horsepower. Conversely, a water velocity that is too low will result in a pipe that is larger than necessary.

Assume, for example, a load of 382,000 lb per hr and the run, before the first take-off, is 400 feet long. What size should the HTW supply and return mains be?

The pipe sizing chart shows that for a load of 382,000 lb per hr, a 6-in. pipe will have a water velocity of 9 ft per sec and a head loss of 2.91 ft per 100 ft.

For comparison, a 5-in. pipe shows a velocity of 14 ft per sec and a head loss of 7.45 ft per 100 ft. An 8-in. pipe shows a velocity of 5 ft per sec and a head loss of 0.69 ft per 100 ft.

Since the velocity and head loss values indicated for a 6-in. pipe are within good design limits, the 6-in. pipe should be specified.

Piping Details

Chapter 8, Steam Distribution, covers piping primarily for steam systems. However, the discussion there of conduits applies equally to HTW systems.

Access Chambers

Long systems of distribution (other than those between nearby buildings) require access chambers to service control valves, expansion joints and pump.

These chambers are constructed of reinforced concrete or steel, and should be drained and ventilated as required.

Concrete access chambers should be constructed of 2500 psi reinforced concrete, not less than 6 in. thick. The thickness of the concrete walls, floor and roof are determined by conditions at the site. The bottom and sides should be formed by monolithic pours and the joint between the bottom and side walls should be waterproof.

Exterior surfaces should be made watertight with a primer, bitumastic coating and membrane waterproofing. The joint between the chamber and the concrete, tile, or cement asbestos conduit should be made tight with bitumastic coating and membrane.

When steel conduit is used, a steel leak plate welded to the conduit should be embedded in the access chamber sidewall by setting the conduit in place before pouring the access chamber wall.

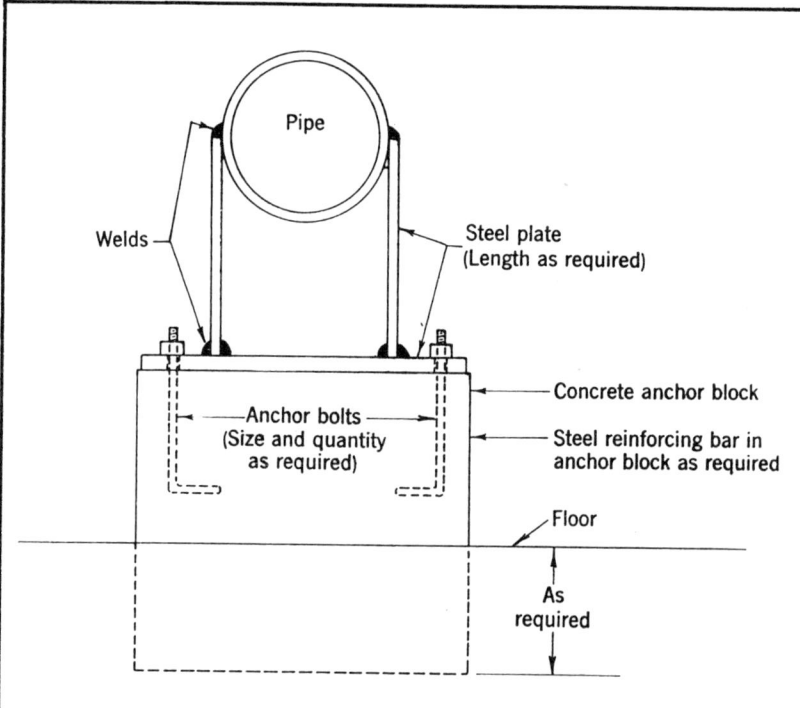

Fig. 1. Floor anchor for large pipes. Anchor, constructed of steel plate, is bolted to a concrete pier. This type of anchor is well suited to pipe lines running close to the floor.

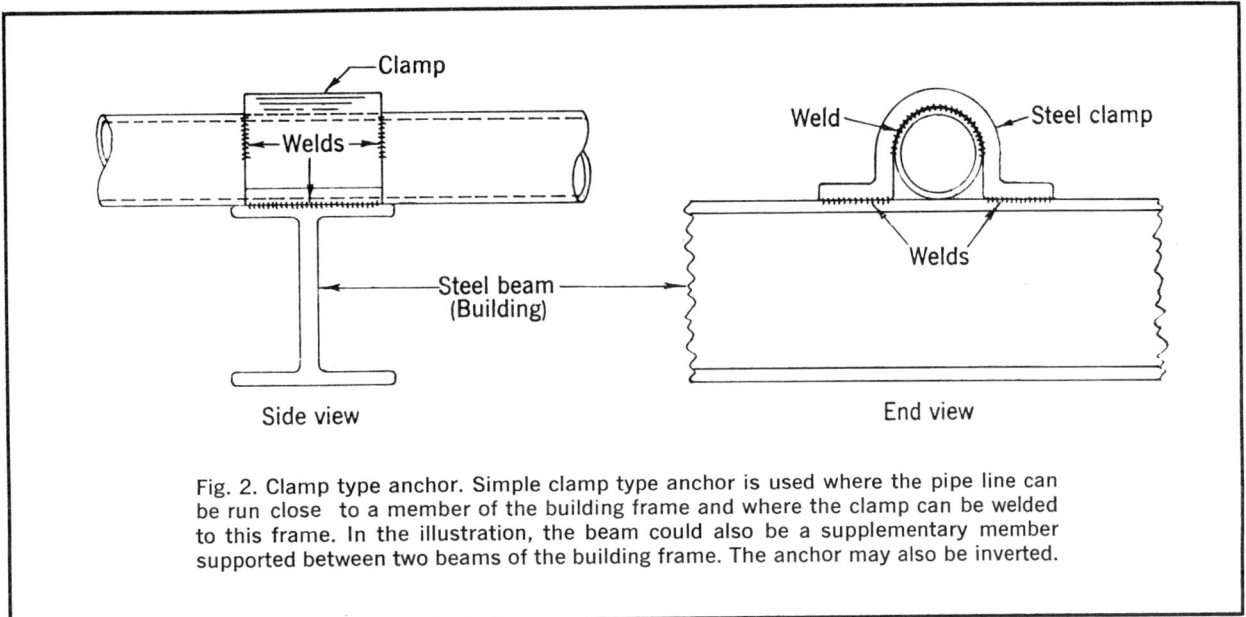

Fig. 2. Clamp type anchor. Simple clamp type anchor is used where the pipe line can be run close to a member of the building frame and where the clamp can be welded to this frame. In the illustration, the beam could also be a supplementary member supported between two beams of the building frame. The anchor may also be inverted.

Piping, expansion joints and valves should be supported upon structural steel frames, built into the walls or floor, as required.

If anchors are to be installed in the chamber, they should be set in the sidewall or floor forms before pouring.

The roof of the access chamber should terminate, if possible, about 12 in. below grade, and only that section supporting the manhole frame and cover should rise to grade. Esthetically, this is the optimum arrangement.

Manhole frames and covers should be sized to allow equipment to be brought into and removed from the chamber; round manholes should be large enough to accept replaceable equipment and never less than 27-in. inside diameter. They should be of the walkway or roadway type, depending on the conditions; covers should always be gasketed and fastened with bronze bolts.

Ventilation of the access chamber is important. Usually, two 8-in. dia pipes with goosenecks

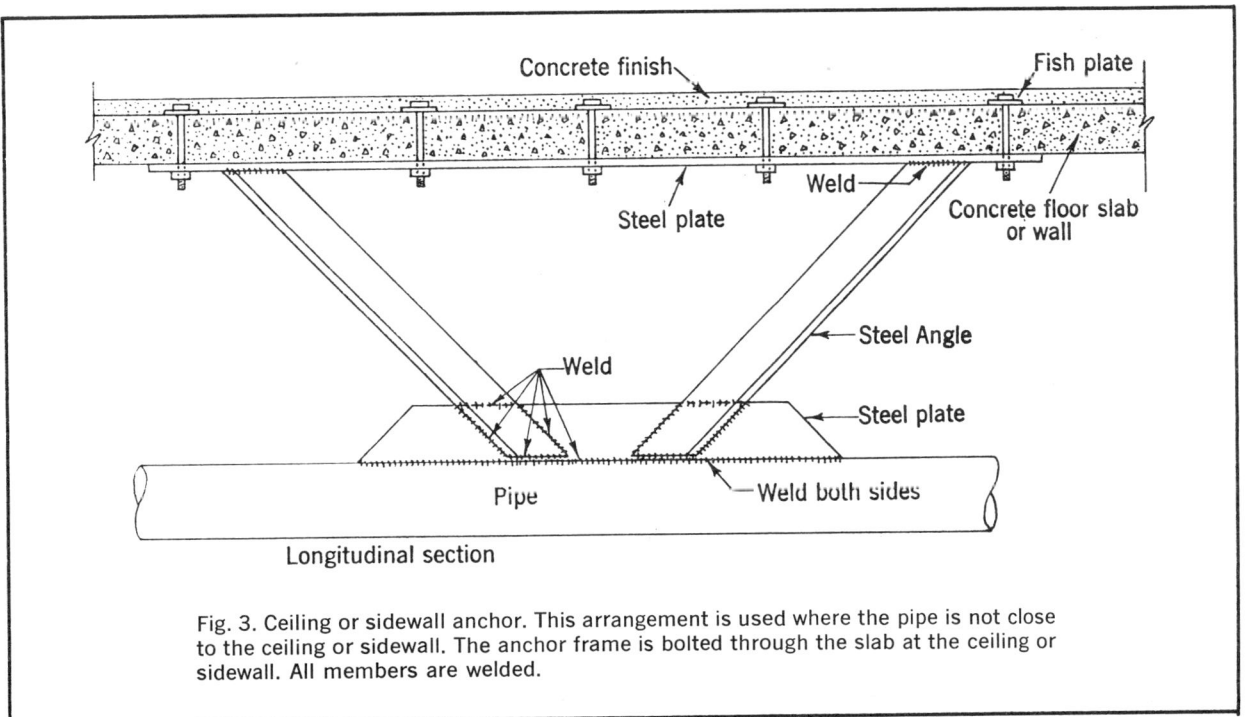

Fig. 3. Ceiling or sidewall anchor. This arrangement is used where the pipe is not close to the ceiling or sidewall. The anchor frame is bolted through the slab at the ceiling or sidewall. All members are welded.

above grade, set in opposite corners of the chamber, with one starting 12 in. from the floor, and the other at the inside roof level, provide adequate ventilation.

If necessary, the chamber should be provided with a sump and pump. If the area is wet, drain tile should be set around the access chamber outside at the lowest level, and run to a sump or collecting basin.

Prefabricated steel access chambers can also be provided, complete with piping and all required components. These access chambers must be constructed and protected from corrosion in the same way as the steel conduit.

Ventilation and removal of mositure for the prefabricated steel type is handled in the same manner as in concrete access chambers.

Anchoring of Pipelines

Proper anchoring of steam, hot condensate, and HTW lines is as important as the line itself. Thermal expansion of pipelines creates an additional force of thousands of pounds pressure and torque. If this thrust is not controlled with adequate anchors and guides, the lines may be pulled from their supports, expansion joints will be broken and expansion loops made useless.

In new buildings the structural engineer must be advised of these additional loads that will be imposed on certain structural members. In existing buildings undergoing renovation, the steel frame must be checked to see if it can withstand the increased stress.

Pipe anchors are made with steel plate and structural steel members. There are several ways to anchor pipelines, depending on the type of building or conduit and on the size of the pipeline. Typical examples are described and shown in Figs. 1 to 6.

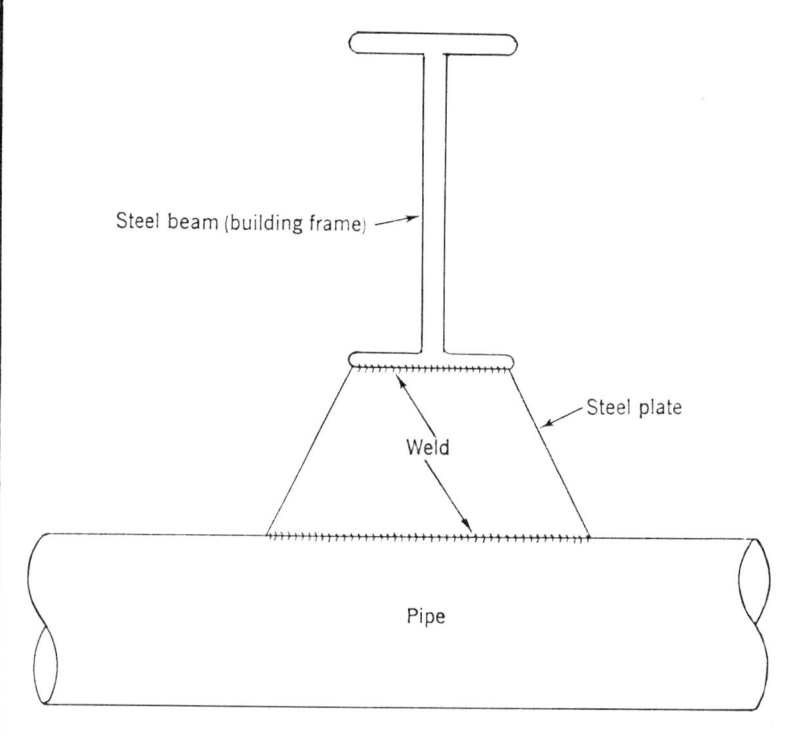

Fig. 4. Ceiling or sidewall anchor. This sketch shows another type of ceiling or sidewall anchor that is used where the pipe is close to the building frame. Composed of a single steel plate welded to the pipe and steel building beam, this arrangement is simple and very effective.

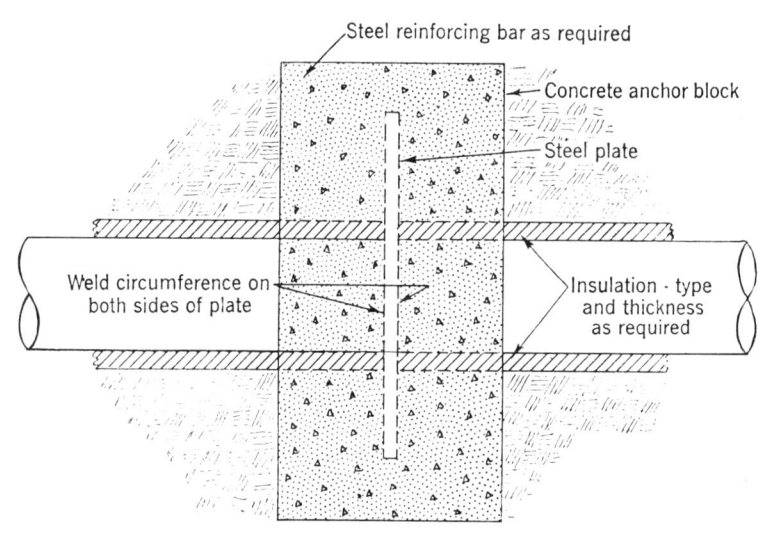

Fig. 5. Underground anchor. How to use a concrete block as an underground anchor. A steel plate is positioned over the pipe and welded to pipe on both sides along the full circumference. Pipe, insulation and plate are then encased in a poured concrete block.

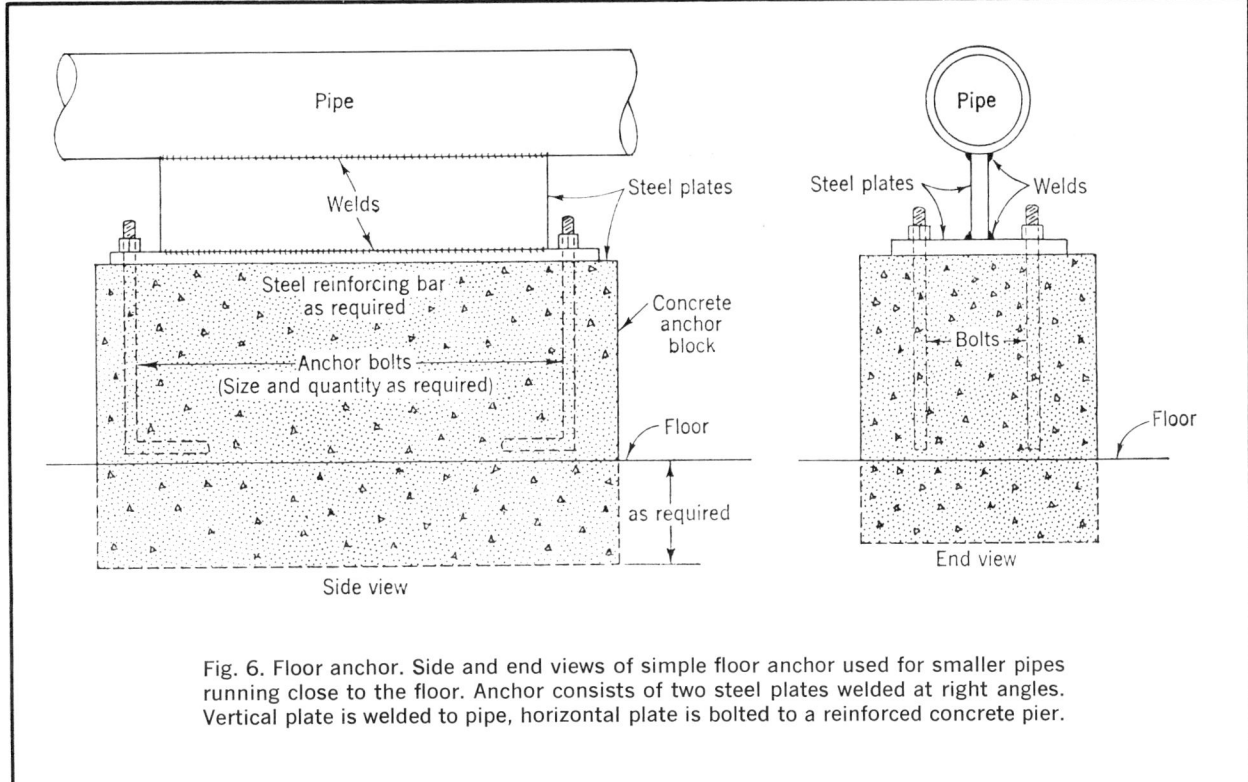

Fig. 6. Floor anchor. Side and end views of simple floor anchor used for smaller pipes running close to the floor. Anchor consists of two steel plates welded at right angles. Vertical plate is welded to pipe, horizontal plate is bolted to a reinforced concrete pier.

In steel frame buildings, pipe anchors are formed by welding steel plate and structural steel members to the pipes and to the steel frame.

In reinforced concrete, anchor bolts must be set in the forms of the walls, beams or floor slabs before pouring the concrete.

Outdoor pipelines above ground are anchored to concrete supports or to steel supporting frames.

Pipelines running underground, in tunnels, access pits or chambers, are anchored to steel members set in the floor or chamber side walls.

In underground steel conduits, the steel plate anchor is welded to the pipe and conduit and extended several inches beyond the conduit sidewall. Conduit and anchor plate are completely encased in concrete to form the anchor block.

In concrete conduit, the steel plate anchor is welded to the pipe and bolted or welded to steel members set in concrete.

Pipe Guides

Pipeline guides serve to keep the lines in position on their supports and to keep the pipes aligned with the expansion joints. This arrangement prevents deformation, especially where an expansion joint is axially loaded.

Pipe guides should be placed in accordance with instructions provided by the expansion joint manufacturer. For example, guides should not be located near expansion loops because this will prevent the loop from expanding properly according to the design calculation.

If it is necessary to place a guide close to an expansion loop, the loop legs must be made longer to compensate for the interference.

To calculate reactions at pipe anchors, consult any standard piping handbook.

Weld Requirements for Anchors

Since pipe anchors are formed from structural steel and steel plate, all welds should be fillet type. The following example shows how to calculate the inches of weld required to anchor a pipe.

Weld value in KIPS per 1/16 in. width per linear in. = 0.6d where d = 1/16 in. width of weld per 1000 lb
Given:
Reaction at pipe anchor = 12,600 lb.
Fillet weld = 3/8 in. wide
Then:
3/8 in. = 6/16 in.
0.6 x 6 = 3.6 KIPS or 3600 lb per linear in. of weld
Therefore, weld requirement is 12,600/3600 = 3½-in. of 3/8-in. wide fillet.

Expansion Joints

When employing the corrugated type expansion joint the thrust factor must be carefully considered as tremendous loads can be imposed upon anchors. This factor varies according to the size of the joint.

Generally speaking the total

thrust is based on the pressure required to compress the expansion joint plus the pressure of the steam or water in the pipe, multiplied by the cross-sectional area (sq inches) of the corrugation, plus the friction created by the pipe guides.

This information is provided by the manufacturers of expansion joints.

Pipeline Expansion

In this example, the following values are given for a steel pipeline:

Distance between anchor points: 200 ft

Pipeline temp = 420F

Installation temp = 40F

Referring to Table 1

Pipe elongation at 420 F = 3.566 in.

Pipe elongation at 40F = 0.430 in.

3.566 − 0.430 = 3.136 in. per 100 linear ft

Therefore, total expansion for 200 ft pipe = 2 × 3.136 = 6.272 or approx. 6 1/4 in.

Table 1. Thermal Expansion of Pipe, Inches per 100 Feet

Temperature, DegF	Steel Pipe	Wrought Iron Pipe	Temperature, DegF	Steel Pipe	Wrought Iron Pipe
−20	0	0	500	4.296	4.477
0	0.145	0.152	520	4.487	4.677
20	0.293	0.306	540	4.670	4.866
40	0.430	0.465	560	4.860	5.057
60	0.593	0.620	580	5.051	5.268
80	0.725	0.780	600	5.247	5.455
100	0.898	0.939	620	5.437	5.660
120	1.055	1.110	640	5.627	5.850
140	1.209	1.265	660	5.831	6.067
160	1.368	1.427	680	6.020	6.260
180	1.528	1.597	700	6.229	6.481
200	1.691	1.778	720	6.425	6.673
220	1.852	1.936	740	6.635	6.899
240	2.020	2.110	760	6.833	7.100
260	2.183	2.279	780	7.046	7.314
280	2.350	2.465	800	7.250	7.508
300	2.519	2.630	820	7.464	7.757
320	2.690	2.800	840	7.662	7.952
340	2.862	2.988	860	7.888	8.195
360	3.029	3.175	880	8.098	8.400
380	3.211	3.350	900	8.313	8.639
400	3.375	3.521	920	8.545	8.867
420	3.566	3.720	940	8.755	9.089
440	3.740	3.900	960	8.975	9.300
460	3.929	4.096	980	9.196	9.547
480	4.100	4.280	1000	9.421	9.776

Expansion Loops

The expansion loop undoubtedly offers the best means for absorbing pipeline expansion because it requires no maintenance and will last as long as the pipeline. However, it is not always the most economical means of absorbing expansion, so that its use must be weighed against other types of expansion joints.

For problems of expansion other than the simple pipe loop, it is suggested that the designer consult a text or handbook on the subject. However, for conditions not requiring rigorous anslysis, it is permissible to design expansion loops by means of the following equation, which the author has successfully used for temperatures ranging to 430F. (Pipe heavier than Schedule 80 must not be used. Fiber stress in this equation does not exceed 16,000 lb per sq in.)

$$L = 6.16 \sqrt{(OD)e}$$

where:

L = length of pipe in feet

OD = outside diameter of pipe in inches

e = pipe expansion in inches

This equation can be used for bends with two fittings, regular U-bend and offset U-bend as shown in Fig. 7. The equation gives the length of the pipe, L, that must be used in the U-bend to take up the calculated expansion, e.

The expansion loop may be cold sprung for one half of its calculated expansion to reduce the stress. Cold springing is accomplished by jacking the loop apart for one half the calculated amount of expansion and then installing the anchors.

Calculating the Expansion Loop

Calculated pipeline expansion = 6 in.

Size of steam pipe = 8 in. Schedule 40

Outside diameter of pipe = 8.75 in.

Using the previous equation,

$$L = 6.16\sqrt{OD \times e}$$
$$= 6.16\sqrt{8.75 \times 6} = 44.66 \text{ ft}$$

Assume that long turn weld fitting will be used. Dimension A must always be greater than Dimension B, since these legs of the bend are the spring pieces.

In order to fit a 3-in. condensate return line loop inside the new loop, it is necessary to make dimension B 5 ft.

Total length of pipe = 44 ft, 8 in.

44 ft, 8 in. − 5 ft = 39 ft, 8 in.

This dimension divided by 2 produces a loop with two 19 ft, 10 in. long legs.

The loop, before anchoring, is then cold sprung 3 in., which is 50% of the calculated expansion.

Pipe Installation in Buildings and Above Ground

Installation of high pressure steam (and HTW) piping within a structure is the least costly of all types of installation.

The frames and floor slabs of the building are used to attach the pipe hangers, which support the piping. Pipe anchors are also attached to the building construction.

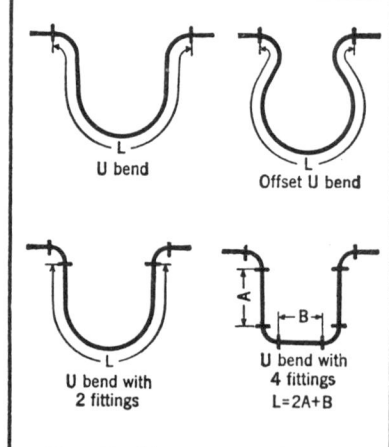

Fig. 7. Measurement of L (length of pipe) in various types of pipe bends. See example in text. Reprinted by permission from ASHRAE Guide and Data Book 1967.

However, it is necessary to inform the structural designer of the loads that are to be imposed upon the frames and floor slabs so that he may take these loads into consideration when designing the structure.

Spring type hangers should be used, particularly where expansion causes pipe movement. Vibration isolators should also be used, where required.

Generally, the suspension of piping poses no problem, but thrust from expansion loops and axially loaded corrugated type expansion joints result in many tons of pressure. Under these conditions, it is sometimes advisable to use either the slip, universal or the ball type of expansion joint in which such thrusts are minimal.

Ball slip and universal type joints also work well in existing structures where additional loadings cannot be imposed upon the building construction.

At industrial sites, it is sometimes necessary to run distribution piping on concrete and steel supports just above the ground level or support it from structural steel frames above the ground. If long distances have to be bridged, the pipeline may be hung from a catenary cable suspended between structural steel towers.

For suspension between buildings, built-up box girders, with their ends resting upon the building side walls, are sometimes used to support the pipelines.

Pipe Insulation

All pipe, valves, fittings and flanges must be insulated. Pipe insulation, today, is generally supplied in 3 ft long sections, completely prefabricated from various insulating materials, and covered with a waterproof, vermin proof jacket (aluminum jackets are also available).

In areas where the insulation may become wet, calcium silicate should be used for high temperatures, and *Foamglas* (a trade name of PPG Industries, Inc.) may be used for low temperatures.

Pipe insulation and jackets used in buildings and above ground should be fire retardant.

Manufacturers' catalogs provide all necessary information for selecting and specifying insulation. If the designer has a problem, manufacturers' sales engineers are helpful and cooperative.

The following table of basic insulating material is set down as an overall guide for the designer.

Calcium silicate	to 1800 F
Solid asbestos	to 1200 F
Glass fiber	to 1000 F
85% magnesia	to 600 F

11

Forced and Induced Draft Fans

In steam generators, the flow of air and combustion gases is confined to ducts, flues and stacks. To supply the correct amount of air for combustion and to remove gases from the generator, this flow must be established and controlled. This can be done by using a stack or mechanical fans. Either the stack or fans or a combination of both produces a difference in pressure, causing gas flow through the boiler unit.

The Forced Draft Fan

A forced draft fan, supplying combustion air to a steam generator, must be of rugged design because it has to operate continuously for months at a time. The motor, housing and bearings should be of the best materials available. The fan wheel must be properly designed and balanced to avoid collecting dirt, which will disturb the balance. The fan must be efficient under a wide range of load conditions and outputs.

Pressure produced by the fan should vary uniformly with output, along a smooth curve within the capacity range. If the fan is driven by an electric motor, it should have self-limiting load characteristics, so that the motor cannot be overloaded. The best fan arrangement for this service is the backward-curved, centrifugal design. This type of fan may be operated at high speeds, which helps to reduce fan size.

Air Control

The simplest and undoubtedly the most practical and economical method of controlling supply air to the fan, is by the multiple inlet vane arrangement. If it is necessary to provide tighter control over the load range, a 2-speed motor may be provided.

Fan Drives

Forced draft fans are driven by either electric motor or steam turbine. When plant heat balance permits, a steam turbine may be used, the exhaust steam being used for feedwater, space or process heating. For the electric motor drive, a squirrel-cage induction motor is suitable because it is relatively inexpensive, reliable and high in efficiency over a wide load range. The steam turbine is more expensive than the electric motor, but highly efficient and more economical to operate.

Fan Selection

The forced draft fan may be used with negatively or positively pressurized furnaces. It may either be required to positively pressurize

56

Forced draft fan. Courtesy of Buffalo Forge Co.

Large, induced draft fan. Courtesy of Buffalo Forge Co.

Example of a Forced Draft Fan Selection (for a Steam Generator Burning No. 6 Fuel Oil, Having an Air Heater)

1. Weight of air to be handled, lb per hour

Theoretical air for combustion (assumed)	200,000
Add 15% excess air	30,000
Total air required for combustion	230,000
Leakage at air heater	10,000
Add 15% safety factor	34,500
Total air for which to size fan	274,500

2. Static pressure requirements, in. of water

Air heater resistance	3.00
Duct resistance	1.00
Resistance of burners and windbox	6.50
Total net pressure required	10.50
Add 25% safety factor	2.63
Static pressure for which to size fan	13.13

3. Temperature of air to be handled, F

Inlet air temperature at fan	80
Add 25F safety factor	25
Operating temperature for which to size fan	105

4. Volume of air to be handled, cfm

$$cfm = \frac{274,500}{0.07353 \times 60} = \qquad 62,300$$

where 0.07353 is the specific weight of air at 80F, lb per cu ft

5. Electric motor requirements, hp

$$\text{Air horsepower} = \frac{62.4/12}{33,000} \times cfm \times \text{static pressure}$$

$$= 0.0001573 \times 62,300 \times 13.13 = \qquad 128.5$$

hp required at 80% efficiency   161.0

The motor should be selected for the nearest size above the non-overloading power of the fan at rated speed. At altitudes other than sea level, cfm and static pressure values should be corrected.

the furnace and gas passes of a steam generator or used in a stoker fired furnace in addition to an induced draft fan or stack that removes products of combustion from the steam generator. It must have enough static pressure to overcome the resistance of coal burning grates and fuel beds, air heaters, ducts, windboxes, burners, registers, etc.

The Induced Draft Fan

The induced draft fan should have the same rugged characteristics as the forced draft fan, and then some, because it must operate under elevated temperatures and handle dirty corrosive and erosive gases. The fan must be able to overcome draft losses over the heating surface and through the gas passes and ducts, then discharge the gas into a stub or tall stack.

Fan Requirements

Fan blading is either forward curved, backward curved or flat. For very dirty or erosive gases the flat blade, operating at low speed, gives the best results. Wearing strips may be also installed on the blading for longer life. Bearings may be either water cooled or air cooled, depending upon gas temperatures and design.

Control of gas flow to the fan is by an automatically controlled uptake damper. Drives for the induced draft fan are the same as for the forced draft unit.

Example of Induced Draft Fan Selection for a Steam Generator Using No. 6 Fuel Oil with Air Heater, Superheater

1. Weight of gas to be handled, lb per hour

Theoretical air for combustion (assumed)	200,000
Add 15% excess air	30,000
Leakage through furnace and boiler	15,400
Leakage, air to gas in air heater	3,500
Fuel (assume no ash) in flue gases	15,000
Total weight	263,900
Add 15% safety factor	39,500
Total air for which to size fan	303,400

2. Static pressure requirements, in. of water

Furnace resistance	0.10
Boiler and superheater resistance	4.80
Air heater resistance	2.30
Duct resistance	0.35
Total resistance	7.55
Add 25% safety factor	1.88
Static pressure for which to size fan	9.43

3. Temperature of gas to be handled, F

Temperature of gas leaving air heater	318.0
Add 10% safety factor	31.8
Temperature of gas at which to size fan	349.8

4. Volume of air to be handled, cfm

$$\text{cfm} = \frac{303,400}{0.04901 \times 60} = \qquad\qquad 103,000$$

where 0.04901 is the specific weight of flue gas at 350F, lb per cu ft

5. Electric motor requirements

$$\text{Air horsepower} = \frac{62.4/12}{33,000} \times \text{cfm} \times \text{static pressure}$$

$$= 0.0001573 \times 103,000 \times 9.43 = \qquad\qquad 153$$

hp required at 75% efficiency   204

6. Motor selection

Proper selection of the fan motor is of the utmost importance. It is doubtful that the next largest motor size above the calculated horsepower input would be adequate. One reason is that the design operating temperature is 350F, but when the fan is started up it will be handling air at boiler room temperature and density, which is much heavier than its actual operating temperature and density. This would cause the motor to overload and its protective circuit breaker to drop out.

Assuming a boiler room temperature of 90F, and a calculated operating temperature of 350F, the following equation will indicate a multiplier for the calculated horsepower input under these conditions.

$$\frac{460 + 350}{460 + 90} = \frac{810}{550} = 1.47$$

$$204 \text{ hp} \times 1.47 = 300 \text{ hp}$$

By this calculation the nearest motor size above the indicated horsepower would have to be selected, but this size would not be economical, causing additional problems. There is no hard and fast rule for the final motor selection. The designer must consult with the fan manufacturer, who is most familiar with characteristics of the fan, in order to work out a safe and economical motor size. At altitudes other than sea level, cfm and static pressure values should be corrected.

12

Combustion Controls and Instrumentation

Proper application of a control system depends on the functions it can perform. As much of the operator's responsibility lies in control and manipulation of the fuel burning equipment, opportunity for significant errors exist. A good control system will perform these functions in an approved and uniform manner, increasing safety and reliability of operation.

COMBUSTION CONTROLS

An operator, working manually, cannot continuously control combustion in the modern steam generator as easily and efficiently as automatic equipment. The function of an automatic combustion control system is to control and maintain the fuel-air ratio at optimum and to maintain the design steam pressure constant. The automatic steam generator makes use of this system mandatory.

There are two basic types of combustion control systems—positioning and metering—that are either electrically or pneumatically operated. The metering system contains refinements that the positioning system does not, but both have their place. If plant air is not available, a small air compressor, supplying compressed air at 50

psig, must be furnished for the pneumatic system.

A positioning system is generally designed to operate on a change in steam pressure that produces proportional changes in fuel and air feed. This type of system, because of its simplicity, cannot maintain optimum fuel-air ratio over the full load range.

In the metering system, flows are measured and balanced against signals for amounts of steam, which makes it possible to maintain optimum fuel-air ratios over a wide load range.

Positioning Type Controls

Components of a positioning type combustion control are:

1. Master steam pressure controller;
2. Fuel control valve with characterized cam for gas and oil or feed control drive for coal;
3. Combustion air fan inlet vanes operator.

The master pressure controller is set to maintain steam pressure at a predetermined set point. Change in pressure from that set point is sensed by the master that changes the position of the fuel control valve, or coal feed control

drive, and of the combustion air fan inlet vanes simultaneously, according to a preset relationship.

Metering Type Controls

In addition to the three main components of a positioning system of control, a fuel measuring device is added in metering type controls that will measure fuel flow and transmit a loading similar to the measured flow. Another device similar to this one transmits an air flow signal. Coupled with these two units, is a fuel-air ratio relay.

In operation, a master pressure controller, set to maintain constant steam pressure acts to move in parallel both fuel control valve (or coal feed control drive) and combustion air fan inlet vanes on sensing a pressure change from the set point. Measured changes in fuel flow and air flow are transmitted to the fuel-air ratio relay that moves to adjust the inlet vanes operator to preserve the desired ratio of fuel and air for the most efficient combustion over the operating range.

This type of system is capable of adjustment of the fuel-air ratio, but is desirable where fuel pressure may vary with load.

Table 1. Selection Chart for Combustion Control Systems

Type of Fuel	Type of control system	Steam generator capacity ranges, lb Steam per hr
Coal	Modulating positioning	to 60,000
	Full metering (steam flow/air flow)	40,000—80,000
	Full metering with oxygen compensation	above 80,000
No. 6 fuel oil	Modulating positioning	to 25,000
	Full metering	25,000—100,000
	Full metering with oxygen compensation	above 100,000
Gas	Modulating positioning	to 25,000
	Full metering	25,000—100,000
	Full metering with oxygen compensation	above 100,000

A combustion control panel with instrumentation for two steam generators. Courtesy of Bailey Meter Co.

Which Type to Select

It can be seen that the metering system of combustion control is a refinement of the positioning system, which may be further refined by additional devices to provide a more efficient system of control. Experience has shown that a good quality positioning system of control has been very satisfactory with steam generators in capacities to 25,000 lb steam per hour. Above this, some type of metering control is necessary for good efficiency.

This is the point at which the designer sits down with control and steam generator manufacturers and decides which control arrangement will provide the best efficiency at the lowest cost for the job to be done. Based upon the capacity of the steam generator and the type of fuel to be burned, Table 1 is a guide to combustion control system selection.

SAFETY CONTROLS

All automatic steam generators fired with gas and oil should be equipped with safety controls. Factory Insurance Association (FIA) sets standards that provide the ultimate in safety protection for these controls.

Gas- and oil-fired steam gener-ators are specified as it is not possible to provide a coal-fired steam generator with much more than a steam over-pressure switch and a low water cut-off, due to the type of fuel used.

All safety controls should be wired for 120 volts, single phase, grounded, with switching devices wired in the hot leg of the circuit.

The following safety controls, which shut off burners, should be provided for a steam generator burning gas and No. 6 fuel oil:

1. High steam pressure switch;
2. Flame failure safeguard and programming control;
3. Low water cut-off (2)—1 float and 1 probe type;
4. Minimum air-flow switch;
5. Atomizing, air-to-steam pressure switch;
6. Low oil temperature switch;
7. Low oil pressure switch;
8. High gas pressure switch;
9. Low gas pressure switch.

These controls should have an annunciating system of lights and an alarm horn. Each light should have a silencing switch for the alarm horn. When the circuit for the light involved has been cleared, the light will go out.

Table 2 (on page 62) indicates complete FIA requirements for burner safety control systems.

Flame Safeguard Control and Programmer

The flame safeguard control and programmer is a very unique device without which no automatic or semi-automatic steam generator would be safe. This device, through its lead sulphide or ultraviolet sensors, sees the pilot and main flames continuously and, upon their failure, closes the fuel supply valve, which must be reset manually before the burner can be relit. It also automatically programs burner operation by initiating and controlling the following operations:

1. Pre-purge (forced draft fan and/or induced draft fan);
2. Trial for pilot flame (10 seconds);
3. Trial for main flame (gas—15 seconds, oil—30 seconds);
4. Monitoring of main flame during burner firing period;
5. Post-purge (forced draft fan and/or induced draft fan).

In very large units some steam generator manufacturers question the use of a post-purge cycle. They believe that a furnace explosion can take place by mixing oxygen with hot, unburned gases subject to ignition from hot refractory.

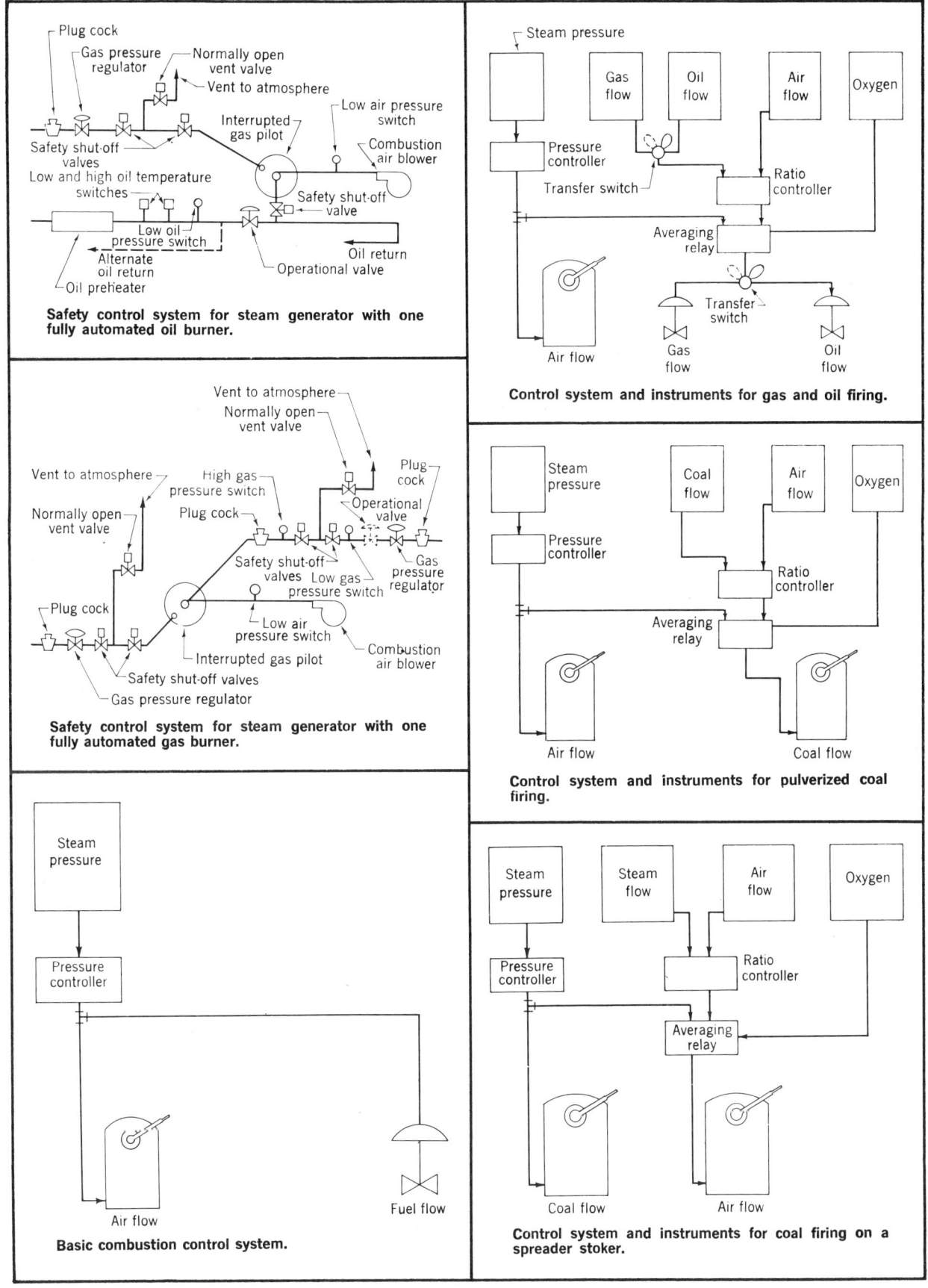

Safety control system for steam generator with one fully automated oil burner.

Control system and instruments for gas and oil firing.

Safety control system for steam generator with one fully automated gas burner.

Control system and instruments for pulverized coal firing.

Basic combustion control system.

Control system and instruments for coal firing on a spreader stoker.

Table 2. FIA Requirements Checklist for Burner Safety Control Systems*

	Boiler Types, Ignition Types and Fuel				
	Multiple Burner	Single Burner		Single Burner	
	Semiautomatic	Manual and Semiautomatic		Fully Automatic	
	Gas and/or Oil	Gas	Oil	Gas	Oil
I Sequence—Normal					
A. Purge					
1. with cocks closed	X	....	....	....	....
2. at 50% air flow	X	....	....	....	....
3. 4 air changes	X	X	X	X	X
4. 5 minutes minimum	....	....	....	....	....
B. Pilot Proving					
1. Required	X	X	X	X	X
2. Limited trial for ignition	10 sec	15 sec	15 sec	15 sec	15 sec
C. Trial for ignition of main flame	10 sec	15 sec	30 sec	15 sec	30 sec
II Sequence—Flame Failure					
A. Shutdown time	4 sec maximum	2-4 sec	2-4 sec	2-4 sec	2-4 sec
B. One burner still firing, one minute waiting time	X	....	....	....	....
C. Total blackout—repurge	X	X	X	X	X
III Alarm System					
A. Individual burner flame failure alarm	X	X	X	X	X
B. Individual open cock alarm during purge and after flameout	X	....	....	....	....
IV Flame Safeguard Electronic	X	X	X	X	X
V Ignition	Gas-electric fixed pilot	Manual: portable, gas torch Semi-automatic: gas-electric fixed pilot		Gas-electric with interrupted pilot	
VI Valves					
A. Pilot	1 Factory Mutual cock 1 manual reset	2 (solenoid or automatic) and a vent over 120,000 Btuh		2 (solenoid or automatic) and a vent over 120,000 Btuh	
B. Main fuel	1 Factory Mutual cock 1 manual reset	1 automatic 1 vent 1 manual reset	1 automatic 1 manual reset	2 automatic 1 vent	1 valve
VII Interlocks					
A. Low water cutoff	....	X	X	X	X
B. High steam pressure	....	X	X	X	X
C. High water temperature (Hot water boiler)	....	X	X	X	X
D. High gas pressure	X	X	X	X	X
E. Low gas pressure	X	X	X	X	X
F. Minimum air-flow switch	X	X	X	X	X
G. Forced and induced fan starter interlock	X	X	X	X	X
H. Low fire start	....	X	X	X	X
I. Stack damper open at start	....	X	X	X	X
J. Low oil pressure	oil only	....	X	....	X
K. Atomizing steam or air pressure	oil only	....	X	....	X
L. Electric proof of burner motor operation	....	....	X	....	X
M. Low oil temperature	oil only	....	X	....	X
N. Purge proving air-flow switch at 50% rate	X	....	....	....	....

*X=Requirement
.... =No recommendation

Control Panels

A dust tight, steel control panel with lock is generally mounted near the front of small steam generators and in the separate main control panel of large units. It contains the forced and induced draft fans, motor starters, guard unit, annunciating lights, sequence operating lights, alarm horn, relays, switches, circuit breakers, etc.

Instruments to inform us of what is occurring in and around the steam generator should be mounted upon a dust tight, lighted, steel control panel, completely separate from the steam generator. This avoids placing instruments and controls upon a console attached to a steam generator and subjecting them to harmful vibration, generally caused by a forced draft fan.

A steel, free standing, dust tight and illuminated control panel must be provided' for mounting combustion controls, safety controls and instrumentation. These instruments and controls may be surface or flush mounted.

The following instruments are those necessary to inform the operator of just what is occurring during plant operation.

1. Steam-flow, air-flow meter (indicating, recording and totalizing);

2. Fuel gas flow meter (recording and totalizing);

3. Feed water flow meter (recording and totalizing);

4. Flue gas temperature recorder;

5. Carbon dioxide, oxygen and combustible gas analyzers;

6. Boiler draft gages (windbox, combustion chamber and last pass);

7. Steam pressure recorder;

8. Feed water temperature recorder;

9. Smoke density recorder (coal and oil only);

10. Fuel oil meter (mechanical at generator). Two required if return flow system is used;

11. Economizer outlet draft or pressure;

12. Economizer inlet-outlet water temperature thermometer;

13. Air preheater outlet draft or pressure;

14. Air preheater inlet and outlet thermometer (combustion air);

15. Steam pressure gage;

16. Feed water pressure gage;

17. Fuel oil pressure gages (as many as required);

18. Fuel gas pressure gages (as many as required);

19. Fuel oil temperature thermometers (as many as required).

Water Treatment

Pure water is necessary to supply makeup for steam generators, for without it they would eventually become piles of corroded steel scrap. Water in its natural state is never pure, not even as rain. It always contains dissolved dust, gases and some suspended matter. In passing over ground surfaces or through rock formations, rain water will pick up various minerals, some soluble, some insoluble.

Two classes of water are:

1. *Surface*—rivers, streams, ponds, brooks, creeks, reservoirs and lakes;

2. *Ground*—waters that seep into the ground, forming deep underground pools that are tapped by shafts or driven wells.

River waters contain a high percentage of minerals, oxygen and silt. Well waters are high in mineral content, carbon dioxide and, sometimes, other gases.

Lake waters are generally turbid, high in oxygen and contain various minerals. The mineral content of river waters varies seasonally, whereas well waters are constant.

Any raw water used to supply makeup for a steam generator must be refined and conditioned before injection.

Raw Water Analysis

In order to determine the type of water treatment to use, it is first necessary to get an analysis of the raw water supply. This analysis can be obtained from the private water company or municipality supplying water to the area or, if a well is driven, the water can be analyzed by a competent laboratory. Often there is more than one source of water supply; in that case, an analysis of each source is necessary.

The quantity of a mineral in water is expressed as grains per U. S. gallon (gpg) or parts per million (ppm). Grains per U. S. gallon multiplied by 17.1 will give parts per million, and parts per million divided by 17.1 will give grains per U. S. gallon.

The term pH is used to express the degree of acidity or alkalinity of water. The number of 7 is the

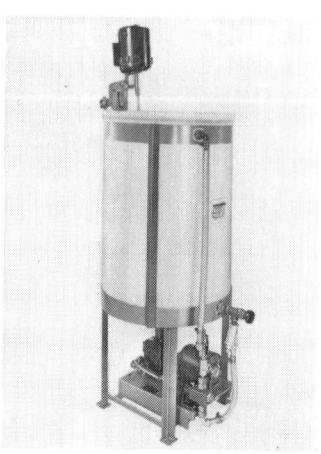

A typical chemical feeding tank and pump with agitator.

A double-unit zeolite softener to remove hardness-producing calcium and magnesium.

neutral point; values above 7 indicate degrees of alkalinity and below 7, degrees of acidity. Water supplies are either acid or alkaline. To find the total hardness of water, which is expressed as calcium carbonate ($CaCO_3$), add the calcium hardness as $CaCO_3$ and the magnesium hardness as $CaCO_3$. The total is expressed as parts per million.

Internal Water Treatment

Water in the steam generator must be chemically adjusted or balanced while evaporation is taking place to prevent scale formation, corrosion, embrittlement and contamination of the steam. Regardless of pretreatment of the feedwater, some internal treatment must be done. The quantity and type of chemicals used will depend on the feedwater analysis and plant operating conditions.

Internal Treatment Equipment

It is of utmost importance that the feeding of chemicals in solution into a steam generator be carefully controlled. This is done by using proportioning pumps with adjustable feed valves. These pumps are of the reciprocating type, driven by small electric motors. Generally, they are of simplex design. These chemical feed units, consisting of a tank and pump, are completely packaged. Tanks should be of stainless steel or heavy plastic.

The average steam plant uses one pump to feed sodium sulfite into the feedwater heater serving the steam generator. However, each generator must have at least one pump to feed the required chemicals to the steam drum. Each steam generator is provided with a chemical feed connection. Dissolved and suspended solids are manually controlled by blowdown from the mud drum.

Another method of controlling blowdown is by using a continuous blowdown system. This method is highly accurate in controlling solids, but expensive due to the fact that heat exchangers and automatic controls are employed. The amount of blowdown must be sufficient to economically justify use of this system.

Water Consultants

Pretreatment of feed water and internal water treatment of steam generators is handled by individual consultants, although a few companies provide both services.

Manufacturers, such as Cochrane Corp. and the Permutit Co. furnish pretreatment equipment and provide complete information on pretreatment of water.

Companies such as the Metropolitan Refining Co. and Water Service Laboratories supply chemicals, equipment, information and service for internal water treatment of steam generators.

Water treatment is a large and complex field. It is a science and the designer is advised to consult with experts in plant design.

Straight sodium zeolite method (shown as NaZ) of pretreating steam generator makeup water is shown above. At left, the dealkalizer arrangement is shown (top), followed by desilicizer (middle) and demineralizer arrangement (bottom).

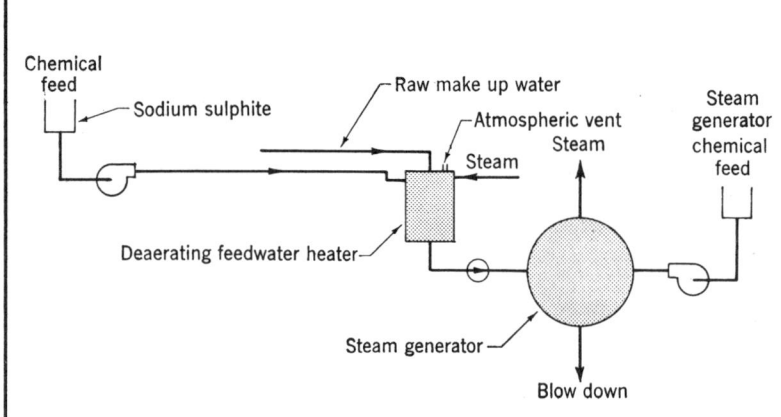

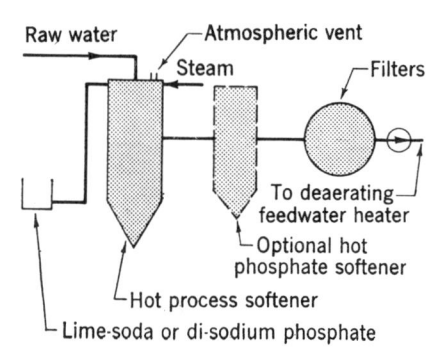

A straight internal water treatment system is shown above. Figure above, right, shows use of chemical precipitation, hot-lime-soda or hot phosphate softeners. A combined chemical precipitation, ion exchange hot lime and zeolite method of water treatment is shown in figure at right.

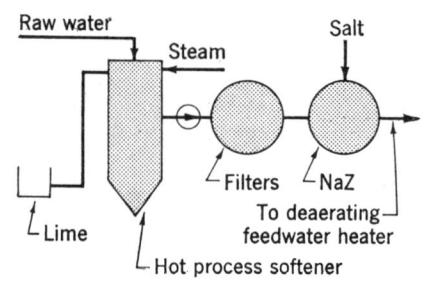

HOW TO DETERMINE MAKEUP WATER TREATMENT—AN EXAMPLE

A steam generator with a capacity of 50,000 lb per hour, operates at 250 psig, having makeup in the area of 10-20% and a hardness below 25 ppm.

With proper blowdown control to regulate total dissolved solids and suspended solids, one may rely on internal treatment only. Above this total hardness, pretreatment as well as internal treatment is required. The feedwater system includes a deaerator to remove oxygen and carbon dioxide.

With naturally soft feedwater, phosphate treatment is generally adequate. Enough phosphate is introduced to reduce hardness of the water in the generator to zero and maintain a phosphate residual of 30-60 ppm. The hydroxide alkalinity may be adjusted with caustic soda to maintain a level from 100-300 ppm. To scavenge any dissolved oxygen in the feedwater, sodium sulfite can be fed into feedwater and a residual maintained at 20-40 ppm in the steam generator.

With a total hardness above 25 ppm, one of the pretreatment water softening systems in Table 1 (as required due to the raw water analysis) must be used to furnish zero-hardness makeup water to the steam generator.

Table 1. Removing Makeup Water Impurities

Impurity	Where Found	Characteristics	How to Remove
Turbidity	Raw water	Suspended matter	Coagulation by precipitation followed by settling and final clarification by gravity or pressure filters
Iron & magnesium salts	Surface and well water	Dissolved salts	Zeolite process
Calcium & magnesium salts (sulphates, bicarbonates, chlorides & nitrates)	All natural waters	Water hardness proportional to content of dissolved salts; capable of forming scale which can cause tube burnout	Sodium or hydrogen zeolite; demineralization; cold lime, lime soda or hot lime soda processes
Sodium salts (calcium, sodium, magnesium)	All natural waters	Dissolved salts; can include sulphates, chlorides & bicarbonates	Combination of hydrogen zeolite and demineralization processes
Silica	All natural waters	Soluble & insoluble; difficult to remove; capable of forming hard scale in steam generators; turbine blades	Insoluble—settling, coagulation & filtration; Soluble—Demineralization; hot, cold lime-soda process; or coagulant (ferric sulphate)
Carbon dioxide	Ground & well waters	Dissolved	Deaerating feedwater heater
	In steam generators	Breakdown of bicarbonates to carbonates to hydroxides	Scavenge—volatile amine
	In condensate return systems	Corrosive to piping	Use a filming amine
Oxygen	All natural waters	Highly corrosive to metal parts	Deaerating feedwater heater will reduce O_2 content to 0.005 cc per liter; add sodium sulphite to complete O_2 removal.

Mineral Analysis of a Lake Water

Mineral	Amount, ppm
Total hardness as $CaCO_3$	46
Calcium hardness as $CaCO_3$	33
Magnesium hardness as $CaCO_3$	13
Alkalinity as $CaCO_3$	46
Sodium potassium as Na	3
Chlorides as Cl	1
Sulfates as SO_4	2
Iron as Fe	0.06
Nitrates as NO_3	0.5
Silica as SiO_2	7

This water is slightly hard. The calcium and magnesium hardness and silica can be removed by demineralization using a strong base anion resin in the anion unit.

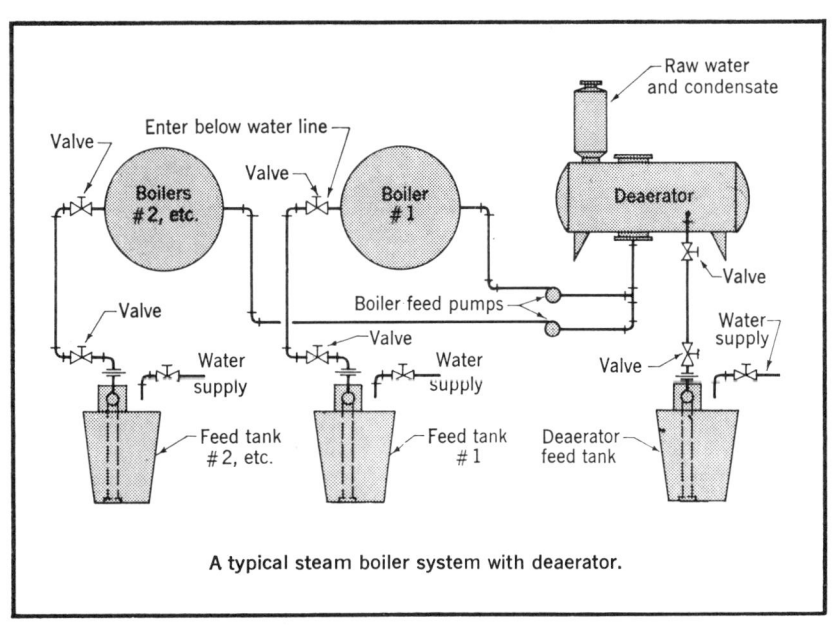

A typical steam boiler system with deaerator.

Boiler-Water Limits for Solids, Alkalinity and Silica

Boiler Pressure at Outlet, psig	Total Solids* ppm	Alkalinity* (Total) ppm	Suspended Solids,* ppm	Silica, ppm
0-300	3500	700	300	125
301-450	3000	600	250	90
451-600	2500	500	150	50
601-750	2000	400	100	35
751-900	1500	300	60	20
901-1000	1250	250	40	8
1001-1500	1000	200	20	2.5
1501-2000	750	150	10	1.0
2001	500	100	5	0.5

*American Boiler and Affiliated Industries Manufacturers Association's maximum limits for boiler-water concentrations in units with a steam drum.
Table used by permission of Power, McGraw-Hill, N. Y.

Chemicals For Steam Generator Internal Water Treatment

Chemical	Purpose	Comment
Sodium hydroxide NaOH (caustic soda)	Increase alkalinity, raise pH, precipitate magnesium	Contains no carbonate, so doesn't promote CO_2 formation in steam.
Sodium carbonate Na_2CO_3 (soda ash)	Increase alkalinity, raise pH, precipitate calcium sulfate as the carbonate	Lower cost, more easily handled than caustic soda. But some carbonate breaks down to release CO_2 with steam.
Sodium phosphates NaH_2PO_4, Na_2HPO_4, Na_3PO_4, $NaPO_3$	Precipitate calcium as hydroxyapatite $[Ca_{10}(OH)_2 (PO_4)_6]$	Alkalinity and resulting pH must be kept high enough for this reaction to take place (pH usually above 10.5).
Sodium aluminate $NaAl_2O_4$	Precipitate calcium, magnesium	Forms a flocculent sludge.
Sodium sulfite Na_2SO_3	Prevent oxygen corrosion	Used to neutralize residual oxygen by forming sodium sulfate. At high temperature and pressures, excess may form H_2S in steam.
Hydrazine hydrate N_2H_4. H_2O (35%) solution)	Prevent oxygen corrosion	Removes residual oxygen to form nitrogen and water. One part of oxygen reacts with three parts of hydrazine (35% solution).
Filming amines Octadecylamine, etc.	Control return-line corrosion by forming a protective film on the metal surfaces	Protects against both oxygen and carbon dioxide attack. Small amounts of continuous feed will maintain the film.
Neutralizing amines Morpholine, cyclohexylamine, benzylamine	Control return-line corrosion by neutralizing CO_2 and adjusting pH of condensate	About 2 ppm of amine is needed for each ppm of carbon dioxide in steam. Keep pH in range of 7.0 to 7.4 or higher.
Sodium nitrate $NaNO_3$	Inhibit caustic embrittlement	Used where the water has embrittling characteristics.
Tannins, starches, glucose and lignin derivatives organic polymers (polyacrylamides)	Prevent feed line deposits, coat scale crystals to produce fluid sludge that won't adhere as readily to boiler heating surfaces	These organics, often called protective colloids, are used with soda ash, phosphate. Also distort scale crystal growth, help inhibit caustic embrittlement.
Seaweed derivatives (Sodium alginate, sodium mannuronate)	Provide a more fluid sludge and minimize carryover	Organics often classed as reactive colloids since they react with calcium, magnesium and absorb scale crystals.
Antifoams (Polyamides, etc.)	Reduce foaming tendency of highly concentrated boiler water	Usually added with other chemicals for scale control and sludge dispersion.
Chelating agents EDTA & NTA	Prevent calcium, magnesium and iron precipitation	Protects against scale formation without forming sludge.

Table used by permission of Power, McGraw-Hill, N. Y.

Coal Handling Systems

Handling and storing of coal in large or small plants is basically the same. Delivery to the plant site is by truck, railroad car or barge; where the small plant may have only indoor storage, the large plant may require extensive outdoor storage.

The coal must ultimately arrive at the stoker or pulverizer hopper. Although types of transfer equipment vary, the end result is the same. Depending on the plant type and size, different arrangements of transfer and storage equipment are employed.

For plants burning up to 10 tons of coal per day, it is possible to handle the coal manually, but not economically. It is therefore necessary to provide completely automatic coal handling equipment. The type of equipment and arrangement used will depend on the plant requirements. When designing the coal handling facility, plant expansion must be considered. Equipment capacity is based on the time allowed for unloading from coal cars, trucks or barges. This usually amounts to 4-6 hours per day.

Coal Storage

Coal may be stored in or out of doors for a considerable length of time, safely, at little or no cost and with minimal loss of heating value. Coal storage is divided into two types. The first is live or active storage under cover, known as in-plant storage, in which the coal is fed directly to the stoker or pulverizer hopper. The second type is known as reserve or inactive storage, in which the coal is stored outdoors and called the stockpile.

In small plants, live storage may be in bins, silos, or bunkers, while in large plants live storage is in a live stockpile. The quantity of stockpiled, live storage coal required varies with type of transportation and geographical location of the plant. With rail de-

Table 1. Factors Influencing Coal Stockpiles for Three Plant Sizes

Factor	Plant No. 1	Plant No. 2	Plant No. 3
	Stockpile Amount, Tons		
Yearly Consumption	5000	15,000	50,000
Reserve Stockpile	1000	3000	10,000
Daily Burn	17	50	170
In-Plant Storage	200	200	510
Active Stockpile	—	350	1200
	Equipment Used for Stacking		
Reserve Stockpile	Rented Conveyor or Bulldozer	Rented Clamshell Crane or Bulldozer	Rented Clamshell Crane or Owned Bulldozer
Active Stockpile	—	Spout from Bucket Elevator or Owned Front-End Loader	Owned Conveyor or Bulldozer

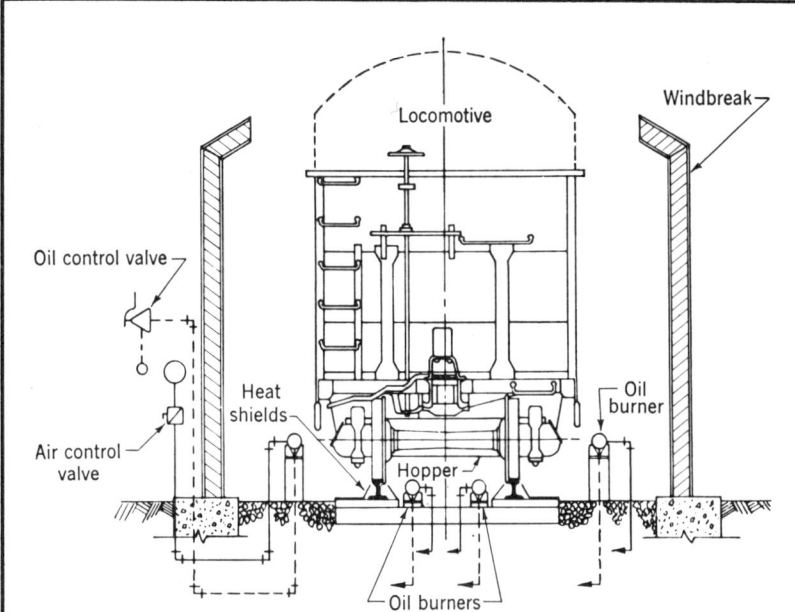

Fig. 1. A coal car thawing arrangement showing wind screens and oil burners.

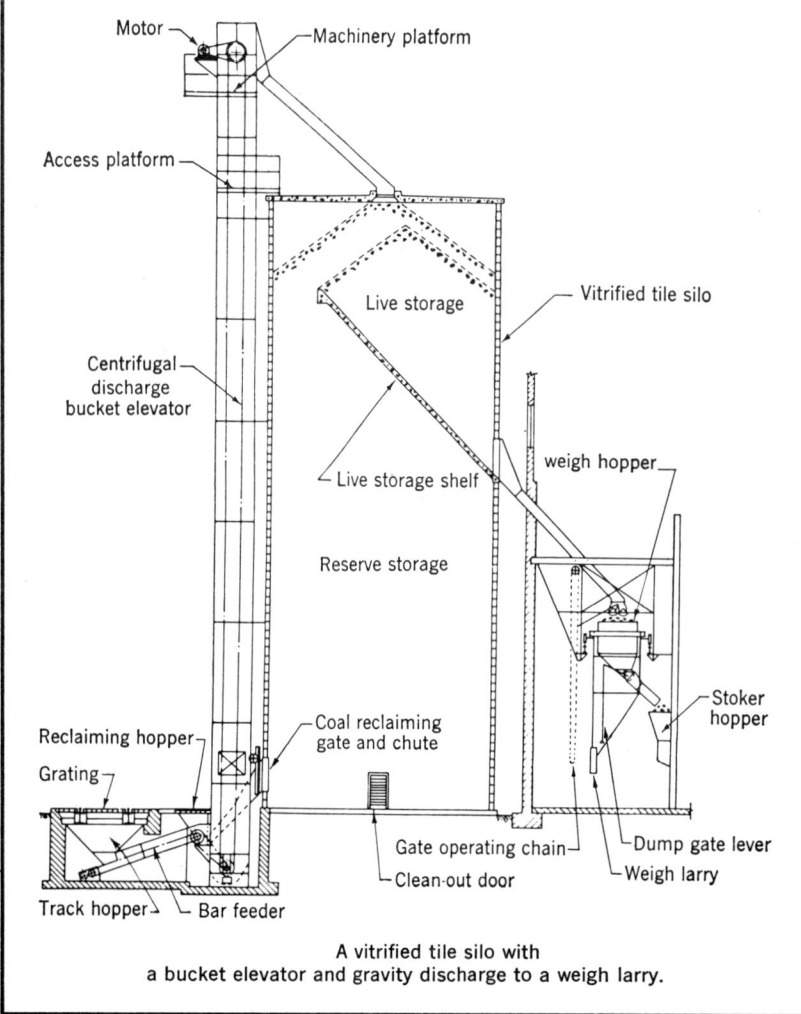

A vitrified tile silo with
a bucket elevator and gravity discharge to a weigh larry.

livery, a 15-day supply should be adequate; for truck delivery, a 10-day supply should suffice.

Minimum Reserve Stockpile

The minimum reserve stockpile should be a 30-day supply based on the maximum coal-burning month or 20% of the total yearly amount burned, whichever is greater. Stockpiles should not be disturbed except for emergencies. In large plants it is not economical to provide large, in-plant storage, therefore a live stockpile, large enough to take care of fuel delivery variations, is provided for.

Types of Storage

Plant size usually determines the type and capacity of storage. Storage quantities vary from the smallest—a small plant without a railroad siding receiving its coal by truck from a dealer who provides storage facilities; to the largest—a large industrial plant that maintains a reserve stockpile and receives its coal by rail, truck or barge.

In-Plant Storage

The small plant usually has a bin capable of storing up to approximately 50 tons in an area that is above freezing temperature. Coal storage facilities of a large plant consist of an overhead bunker, and/or silo, complete with handling equipment that is self-cleaning and designed for recycling coal that stands for more than a month.

Coal is usually stored indoors in straight-sided bunkers, which seem to be the most popular design. These bunkers are constructed of acid-resistant reinforced concrete, steel plate, stainless steel-clad plate or steel plate lined with acid-resistant reforced concrete. The bunkers should have a live capacity of at least 30-72 hours to cover 3-day weekends in which the cost of labor to operate coal handling equipment is prohibitive or personnel not available.

Coal is fed by gravity from an overhead bunker through a short chute to a weigh larry that weighs and feeds the coal to the stoker or pulverizer hopper, or directly to the stoker or pulverizer hopper.

Silos

The outdoor coal silo is generally constructed of tile or reinforced concrete. It provides approximately 15-25% live storage of the total stored. Reserve storage is reclaimed from the bottom by the coal elevator and transported upward to the live storage bin.

Outside Storage

The outdoor site for locating a storage pile should be raised above the surrounding area, well drained, clean and solid. Coals must be compacted to reduce oxidation and deterioration in heating value.

All coals may be compacted with a bulldozer—bituminous in a large continuous pile and subbituminous and lignite in small separate piles. Fire will not develop in a well planned and properly compacted storage pile. Coal may be moved to the elevator hopper by self-powered scrapers, carryalls, bulldozers, draglines or belt conveyors.

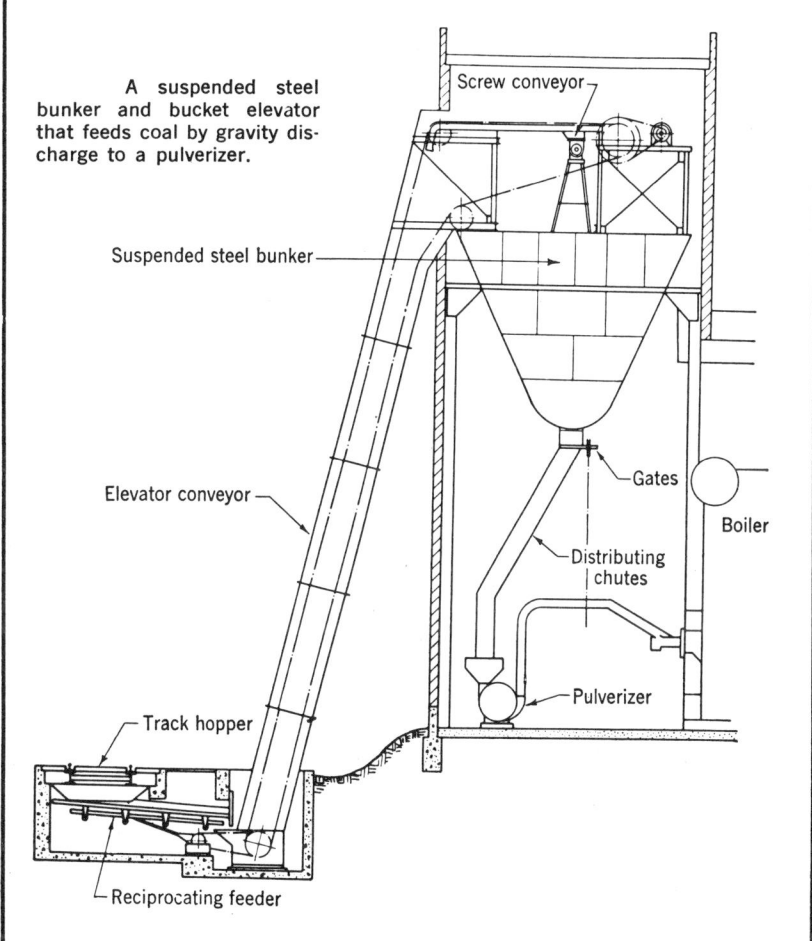

A suspended steel bunker and bucket elevator that feeds coal by gravity discharge to a pulverizer.

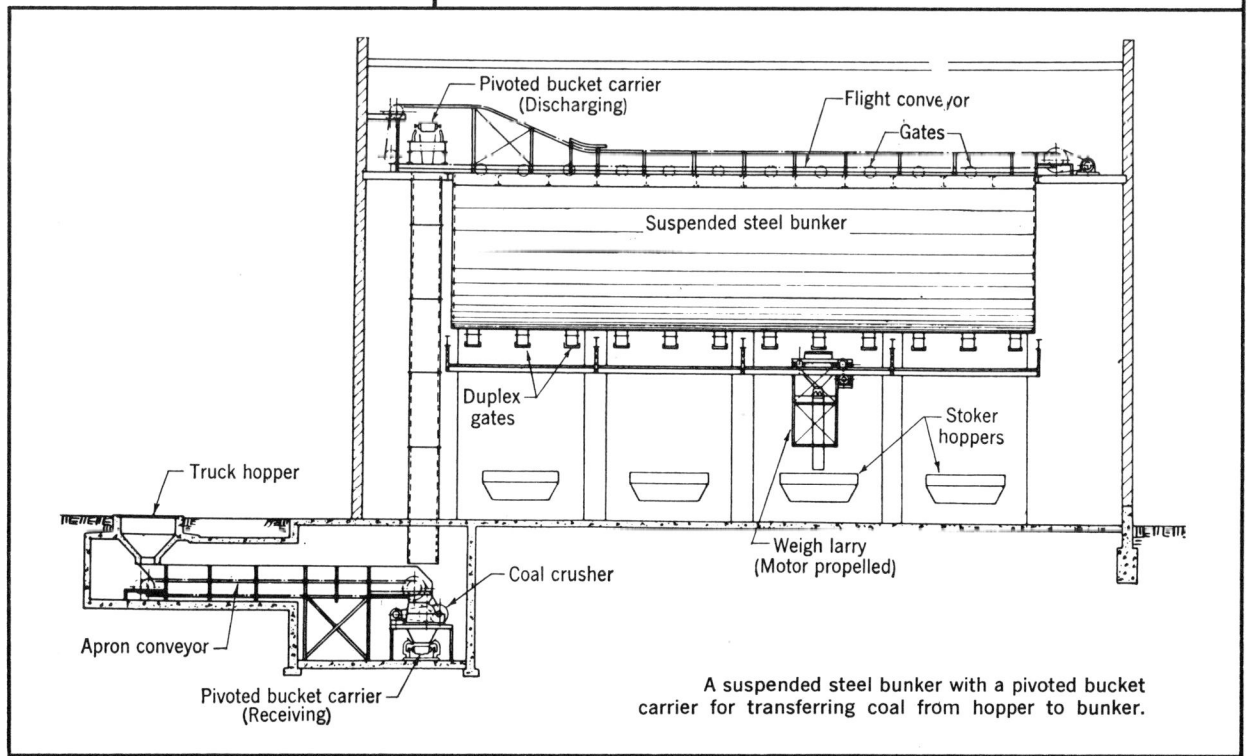

A suspended steel bunker with a pivoted bucket carrier for transferring coal from hopper to bunker.

Stockpile Coal Handling Equipment

Use of modern, mobile equipment to remove coal from a stockpile has come into wide acceptance. This equipment includes bulldozers, tractors, scrapers, carryalls and front-end loaders. For the plant having a stockpile of 10,000 tons or less, a bulldozer that has a practical radius of 300 ft can handle the stockpile, working one shift per day. It is cheaper to rent coal moving equipment for this plant than to own and maintain it. Table 1 indicates storage needs for three sizes of plants with equipment recommended to handle the storage.

Unloading Coal

Coal transported in open railroad cars (as most is) is exposed to every conceivable type of weather. In winter it usually arrives at the plant frozen, unable to flow from the car. Many methods are used to start the coal flowing—striking the car sides with a sledge hammer, breaking the coal from above with a slice bar, as well as car shakeout devices, thawing pits and sheds with oil and gas-fired heaters.

A necessary device for the small plant is the car puller. This is a small fixed electric winch, strategically located along the track, that positions railroad cars over the track hopper and, after dumping, moves them to the far end of the siding. In large plants a bulldozer as well as a car puller is used.

The thawing pit is actually a refractory combustion chamber installed between tracks, having two oil or gas burners. Flames from the burners impinge upon the refractory sides and heat them to incandescence. Radiated heat from the hot refractory then rises up through the car hoppers and thaws the coal. Necessary adjuncts to the thawing pit are steel, transite or aluminum side walls located alongside the tracks, as in Fig. 1, to overcome wind and direct the heat in and around the coal car. Fuel consumption varies between 8 and 15 gal per 50-ton car. Thawing time can range from 20 to 90 minutes.

Thawing sheds that are steam heated and completely enclosed are also used. This arrangement is more costly and thawing time is longer. Sheds must provide adequate headroom for a man standing on top of a coal car. Simple car shakers used in conjunction with these thawing arrangements will help to speed up the process.

For a very large plant, an oil-fired, multiple burner arrangement with windbreaks, such as in Fig. 1, thaws the cars quickly and efficiently.

The Railroad Siding

Plants receiving part or all of their coal by railroad must have a railroad siding large enough to hold an adequate number of filled cars, plus enough space to store several empty cars. The length of the siding depends on the fuel requirements of the plant.

Coal Transport at the Plant

The transport of coal from truck, railroad car or stockpile to the stoker or pulverizer hopper involves a considerable quantity and variety of equipment. Coal usually flows by gravity from the truck or railroad car or is pushed by a bulldozer from the stockpile hopper and falls on an apron feeder, bar feeder or reciprocating feeder that moves it to a crusher. It is then discharged to the boot of the coal elevator. The crusher may also be bypassed and the coal fed directly to the elevator boot.

The elevator, by means of buckets on an endless chain, lifts the coal to its highest elevation and spills it into a silo. From the live storage section, it is chuted to a weigh larry, stoker hopper bucket carrier, screw conveyor or belt conveyor for distribution along the length of the bunker. Inclined belt conveyors are used in large plants for transporting coal from hopper to overhead bunker or silo. Such conveyors must have a slope not more than 18° from horizontal. They are low in maintenance cost.

Manually operated gates in the bottom of the bunker allow coal to be fed by gravity to a weigh larry, where it is weighed and deposited in the burner hopper or through closed chutes with weighing devices to the stoker or pulverizer hoppers. The coal, after reaching the top of the elevator, may also be diverted to a movable chute that, by gravity, places the coal in the live outdoor stockpile. All moving equipment is driven by electric motors.

Pulverized Coal Firing

The pulverized coal burner is used in firing steam generators in capacities of from 100,000 lb per hr to the largest size units built. Skilled workers are required to operate this equipment, in contrast to a stoker that requires less skill to operate. However, the higher operating costs of skilled labor are offset by higher efficiency.

Coals Used

Grindability, rank, moisture, volatile matter and ash are important properties in the selection of coals for pulverizing. Coals having the following limits are required: maximum total moisture* (as fired)—15%; minimum volatile matter (dry basis)—15%; maximum total ash (dry basis)—20%.

Fineness is the most important factor in the burning of coals of various ranks. Table 1 indicates required fineness.

Relative hardness, or grindability, of coal is important. At least two major manufacturers base the output of their pulverizers upon

coal with a grindability of 50-55 (ASTM) and a fineness of 70% through a 200-mesh sieve.

The Pulverizer

Of the many systems to pulverize coal for burning in steam generators, the direct-firing pressure system is undoubtedly the best. This is because it is low in first and operating costs, flexible to control, reliable, compact and safe from fire hazard. The turn-down ratio without changing burners is approximately 3 to 1.

The Babcock and Wilcox Type EL Ball and Race pulverizer is typical of the direct-firing type that operates under pressure. This design has a range of 1½-25 tons of coal per hour, with a very fast response to load change. All coal drying is done within the unit. Adjustment of grinding elements is external, which allows the operator to change settings without shutdown. The pulverizer is operated by electric motors.

Control of the feeder is maintained by sensing the relationship

Table 1. Required Pulverized-Fuel Fineness at Maximum Rating

Per cent through 200 U.S.S. Sieve*

| | ASTM classification of coals by rank | | | | | |
| | Fixed carbon, % | | | Fixed carbon below 69% | | |
Type of Furnace	97.9-86.0 Petroleum coke	85.9-78.0	77.9-69.0	Btu above 13,000	Btu 12,900-11,000	Btu below 11,000
Marine boiler furnace	—	85	80	80	75	—
Water-cooled furnace	80	75	70	70	65**	60**
Cement kiln	90	85	80	80	80	—
Metallurgical	(As determined by process, generally from 80 to 90%.)					

* The 200-mesh screen (sieve) has 200 openings per linear in. or 40,000 openings per sq. in. From U.S. and ASTM sieve series, the nominal aperture for 200 mesh is 0.0029 in. or 0.074 mm. The ASTM designation for 200 mesh is 74 microns.

** Extremely high-ash-content coals will require higher fineness than indicated.

*This limit may be exceeded for lower rank, higher-inherent-moisture-content coals, such as bituminous and lignite.

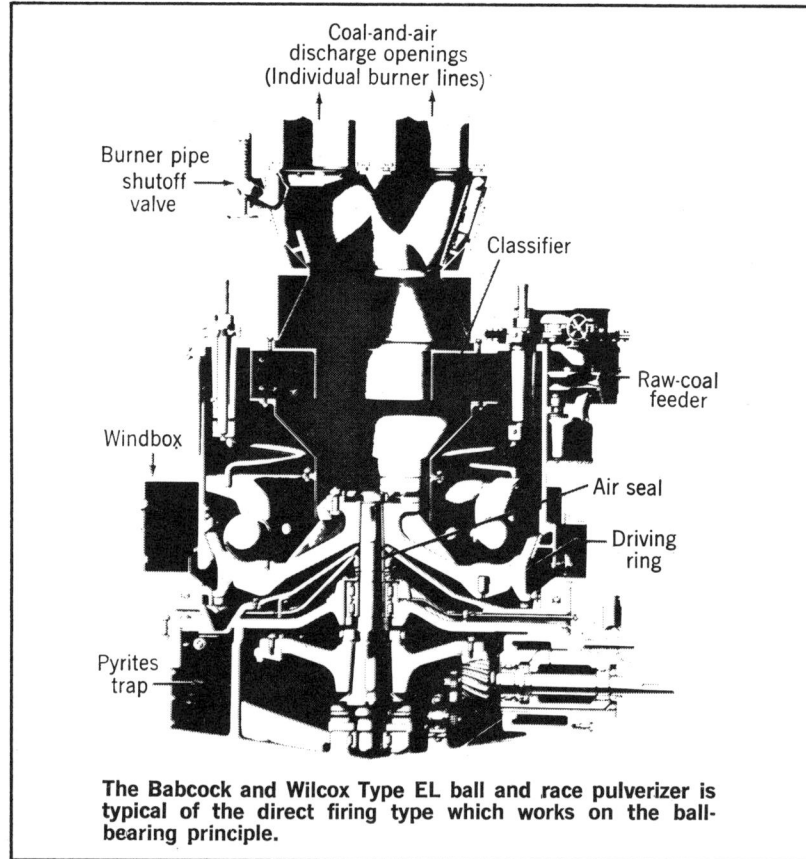

The Babcock and Wilcox Type EL ball and race pulverizer is typical of the direct firing type which works on the ball-bearing principle.

The most frequently used burner for pulverized coal is this circular type one. Courtesy of The Babcock & Wilcox Co.

between pressure drop across the pulverizer and the pressure differential of a pitot tube or orifice in the air supply to the pulverizer. The feeder-controller system operates on this relationship with air flow through the pulverizer varying either manually or by automatic combustion control. The feeder-controller then automatically adjusts coal feed to air flow maintaining the proper fuel-air ratio across the load range. Generally accepted safe values for pulverizer exit fuel-air temperatures are as follows:

Fuel	Exit Temp., F
Lignite	120
High volatile bituminous	150
Low volatile bituminous	150-175
Anthracite	200
Petroleum coke	200-250

Entering air temperature may be as high as 600F. This varies depending on quantity of surface moisture and type of pulverizer. The heated air is furnished by the steam generator air heater. A fan, operating at a positive discharge pressure strong enough to overcome various losses along the line from pulverizer to burner, provides the air to move the pulverized coal, dry it and, at the same time, provide primary air for combustion.

There are many types of coal burners available: the circular burner, multiple-intertube multi-tip burner and the cross-tube burner, to name a few. All fit into different firing design patterns, but the circular type burner seems to be quite popular, and is quite often employed.

The most important characteristic of this burner is that for reliable and efficient performance it requires little attention. It can be equipped to burn gas and oil very efficiently. A single burner has a maximum capacity of 165,000,000 Btuh. This type of burner is generally used on dry-ash-removal type furnaces. It can be used in slag-type furnaces with the burner wall formed entirely of furnace cooling tubes.

Mechanical Coal Stokers

In this age of mechanization it is not practical, economical or efficient to manually fire steam generators with solid fuels. While it is possible to do so with small units, costly labor and poor efficiency have increased use of the automatic stoker, which does the job efficiently and with a minimum of attention.

The mechanical stoker can burn anthracite, bituminous, lignite and coke breeze coals, as well as other fuels such as hogged wood, bark and bagasse.

It is designed to mechanically feed coal in uniform quantities onto a grate within a furnace and to remove ash from the furnace zone. High combustion rates are possible. Continuous firing makes it possible to have good control and high efficiency.

Overfire air jets are necessary to mix gases for efficient combustion. Fly-ash reinjection is used, where warranted, to assist ignition of incoming raw fuel.

The successful stoker installation depends on matching the stoker with type and size of fuel to be burned and capacity required. The steam generator selected must be adaptable to the selected stoker. Stokers are limited in capacity to approximately 400,000 lb steam per hour.

The three main groups of stokers are overfeed, spreader and underfeed. Included in the overfeed are the chain grate, traveling grate, vibrating grate and oscillating grate stokers.

Chain-Grate and Traveling Grate Stokers

Chain and traveling grate stokers have assembled links, grates or keys joined together in endless-belt arrangements that pass over sprockets or return bends at the front and rear of furnaces. Coal, fed from a hopper onto the moving assembly, enters the furnace after passing under an adjustable gate that regulates thickness of the fuel bed. The layer of coal on the grate entering the furnace is heated by radiation from furnace gases and is ignited together with hydrocarbon and other combustible gases that have been driven off by distillation.

The fuel bed continues to burn as it moves along. As combustion progresses, the bed becomes correspondingly thinner. At the far end of travel, ash is discharged over the end of the grate into the ash pit.

A hydraulic drive unit rotates the grate through a ratchet-and-pawl mechanism applied to the front end of the stoker. The power unit may be a constant speed motor or a steam turbine.

The quantity and control of combustion air to the various sections of the stoker is important. The stoker is zoned, or sectionalized and has individual zone dampers to control pressure and quantity of air to various sections as fuel travels along its length. A forced draft fan furnishes the necessary amount of air at the proper static pressure for combustion.

Vibrating-Grate Stoker

The vibrating grate stoker has a grate structure supported upon equally spaced vertical plates that are free to move back and forth. The grate structure is made up of water-cooled tubes arranged longitudinally between a pair of headers. Cast tuyere blocks are fitted on top of the tubes and keyed in position. Coal is hopper fed, with thickness of the fuel bed regulated by an adjustable gate. Vibration is provided by a generator at the front of the stoker beneath the coal hopper. The vibration generator consists of contrarotating mass weights, physically arranged to cancel any motion in a vertical direction. The stoker is driven by a constant speed motor that is op-

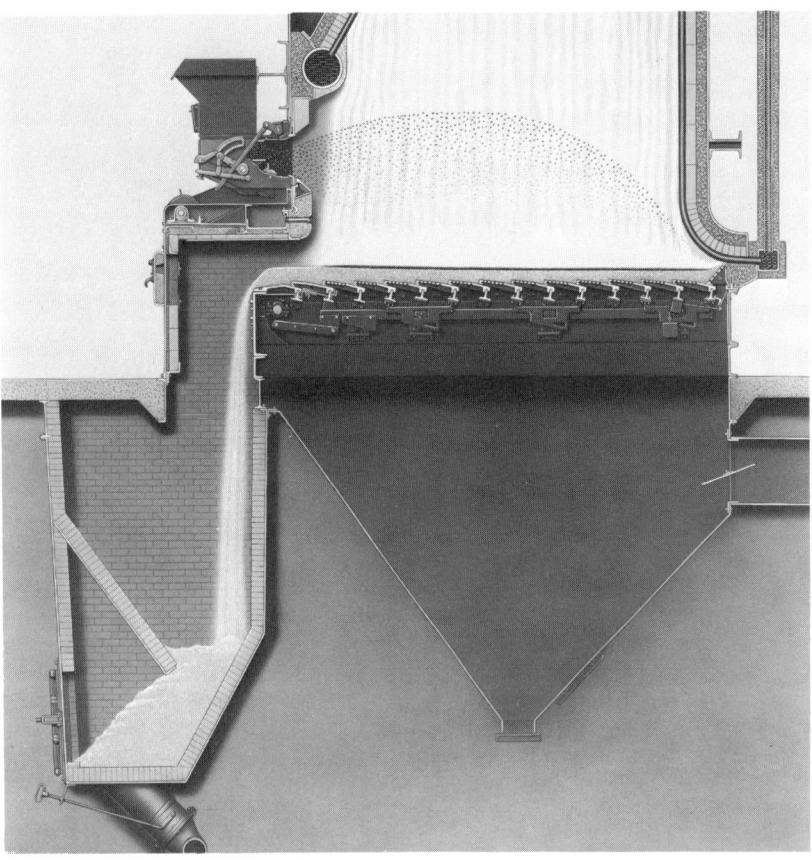

Reciprocating grate spreader stoker with front end discharge. Courtesy of Detroit Stoker Co.

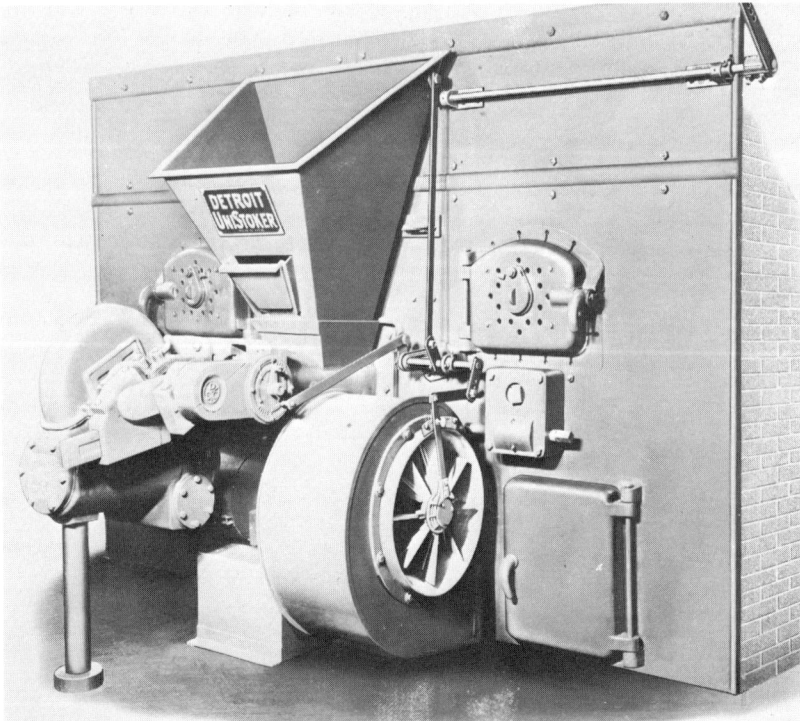

Underfeed stoker with integral mounting of common motor drive for both stoker and draft fan. Courtesy of Detroit Stoker Co.

erated intermittently by a timer control. Grate speed is controlled by frequency of grate vibration.

Spreader Stoker

The spreader stoker uses a combination of suspension burning with a thin, fast-burning fuel bed. It is very sensitive to load fluctuations, as ignition occurs almost simultaneously with an increase in the firing rate.

Grates may be stationary, intermittent dumping or continuously-cleaning (by traveling, reciprocating or vibrating), but it is essential that the grate design incorporate a high-resistance, air-metering principle for best results.

The stoker consists essentially of feeder units arranged to distribute fuel evenly over the grate area. Forced draft, overfire air and combustion control systems are a necessity with this design. Drive arrangements may be mechanical or hydraulic. Combustion air is furnished by a fan to an undergrate, compartmented, plenum chamber for the dumping or stationary grate types, and to one large plenum for traveling grate type units.

Underfeed Stokers

The underfeed stoker is classified in two types—horizontal feed and gravity feed. In the horizontal feed type, coal flow in a longitudinal channel within the furnace, known as a retort, is parallel to the station floor. In the gravity feed type the retort is inclined at an angle of 20° to 25°. The latter type requires a basement or tunnel under the floor for ash disposal, whereas for the horizontal feed type stoker, a shallow depression or pit suffices. Only the multiple-retort gravity type is suitable for the higher capacities within the range of an underfeed stoker.

Side-Ash-Discharge Stokers

In the side-ash-discharge, underfeed stoker, coal is force-fed to the fuel bed in small increments intermittently by a ram or, for very small stokers, continuously by a screw. After a retort is full, the fuel is forced upward and spills over the top on each side to form and to feed the fuel bed. Air is supplied through tuyeres at each side of the retort and through openings in the side grates.

As the coal rises, heat travels downward from the burning fuel above, the volatile gases being distilled off and burned as they pass through the incandescent fuel bed. The rising fuel then ignites upon contact with the actively burning bed. Pressure exerted by incoming raw coal moves the fuel bed gradually over the tuyeres and side grates while burning continues. Combustion is completed by the time the bed reaches the side-dump grates from which the ash is discharged to shallow pits.

Underfeed stoker with separately operated overfeed section and ash discharge plates. Courtesy of Detroit Stoker Co.

Grates—Some side-dump, underfeed stokers rely on pressure from incoming raw fuel to achieve distribution over the side grates. This is satisfactory for single-retort stokers up to a width of 8 ft. For wider side-dump, underfeed stokers, distribution is obtained by reciprocating tuyere blocks between double-retort sections. Other designs have a single center retort and depend upon agitating grates for distribution.

Underfeed stokers for residential use and very small commercial application have round retorts and stationary grates from which ash is removed manually in the form of clinker. Larger side-dump, underfeed stokers are constructed with long center retorts, adjacent to which are tuyere grates. These are followed by side grates and dump grates with air openings that permit a final burnout before ashes are discharged to shallow pits for ultimate cleanout through doors in the stoker front.

Air distribution—Enough combustion air is supplied for the fuel bed conditions from a plenum chamber, or windbox. It is admitted to the fire through the tuyere openings and perforated grates. Manually operated dampers admit air from the central plenum chamber to the ash pits under the side-dump grates, for final burnout before the ash is dumped.

It is quite common for overfire air to be used with underfeed

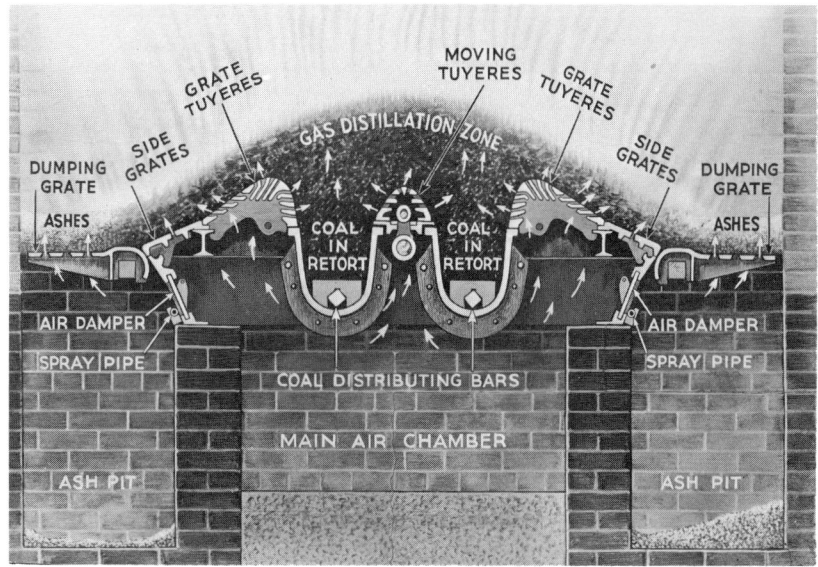
Double retort, horizontal feed type of side-ash-discharge underfeed stoker. Courtesy of Detroit Stoker Co.

stokers to provide some combustion air in the flame zone directly above the active fuel bed. This air may be drawn from the windbox, from a separate overfire-air fan or both. Overfire air is quite effective in preventing smoke, especially at low loads with a lazy fire and on sudden increase of coal feed with the distillation of large quantities of volatile gases.

Rear-Ash-Discharge Stokers

The rear-ash-discharge, gravity-feed type of underfeed stoker is usually longer than the side-ash-discharge, horizontal-feed type and is always designed with multiple retorts. The multiple-retort, gravity-feed stoker, a natural development of principles success-

fully applied in the single retort, consists of a series of inclined single retorts placed side by side with tuyeres between and sloped from front to rear for constant ash discharge.

Each retort is equipped with a round or square primary ram that feeds coal into a ram box at the head of the retort. From this point the fuel bed is moved slowly toward the rear and, at the same time, is forced upward over the bank of tuyeres by secondary pushers or by the moving bottom of the retort.

Most of the air for combustion is admitted through boxes supporting the tuyeres, located between the retorts. An overfeed or cleanup section is provided at the rear end of the bank of tuyeres

Table 1. Stoker Selection Criteria

Type of Stoker		Range of Boiler Capacity to Which Applicable (Mbh)	Fuel Requirements		Maximum Burning[a] Rate (Btu/ft²/hr)	Maximum Amount of Excess Air With Which Complete Combustion Should Be Achieved[b] (%)	
Type of Feed	Type of Grate		Size	Characteristics		At Maximum Load	At Minimum Load (indicated in parenthesis[c])
Underfeed—single and double retort (screw or ram)	Stationary	Up to 5000	2 or 1¼ in. x 0 nut & slack, not to exceed 20% through ¼ in. round-hole screen	Free burning to medium caking (free swelling index less than 5 or 6). Volatile matter moisture-free 22%; min. ash content (dry) 4½%; min. ash-softening temp. 2100°F[d].	300,000	35	60 (at ⅓ load)
	Stationary plus dumping	Up to 10,000	2 or 1¼ in. x 0 nut & slack, not to exceed 20% through ¼ in. round-hole screen	Free burning to medium caking (free swelling index less than 5 or 6). Volatile matter moisture-free 22%; min. ash content (dry) 4½%; min. ash-softening temp. 2100°F[d].	300,000	35	60 (at ⅓ load)
	Undulating plus dumping	Up to 30,000	2 or 1¼ in. x 0 nut & slack, not to exceed 20% through ¼ in. round-hole screen	Free burning to medium caking (free swelling index less than 5 or 6). Volatile matter moisture-free 22%; min. ash content (dry) 4½%; min. ash-softening temp. 2100°F[d].	500,000	30	55 (at ¼ load)
Underfeed—multiple retort	Inclined stationary plus either dumping or undulating, or undulating plus dumping	Over 30,000	2 in. x 0 nut & slack, not to exceed 40% through ¼ in. round-hole screen	Free swelling index less than 6, volatile matter moisture-free 26%, or up to 34% with special design; min. ash content (dry) 4½%; min. ash-softening temp. 2100°F[d].	600,000	25	55 (at ¼ load)
Overfeed (spreader)—underthrow or overthrow rotor or pneumatic	Dumping	15,000 to 50,000	Good mixture of 1½ in. to 1 in. x 0 nut & slack, not to exceed 40% through ¼ in. round-hole screen	Bituminous, subbituminous, or lignite. Min. ash content (dry) 8%[e].	450,000	30	45 (at ¼ load)
	Undulating plus dumping	Over 20,000	Good mixture of 1½ in. to 1 in. x 0 nut & slack, not to exceed 40% through ¼ in. round-hole screen	Bituminous, subbituminous, or lignite. Min. ash content (dry) 8%[e].	600,000	30	45 (at ¼ load)
	Vibrating or pulsating	Over 20,000	Good mixture of 1½ in. to 1 in. x 0 nut & slack, not to exceed 40% through ¼ in. round-hole screen	Bituminous, subbituminous, or lignite. Min. ash content (dry) 8%[e].	600,000	30	45 (at ¼ load)
	Traveling	Over 50,000	Good mixture of 1¼ in. to ¾ in. x 0 nut & slack, not to exceed 40% through ¼ in. round-hole screen	Bituminous, subbituminous, or lignite. Min. ash content (dry) 8%[e].	600,000 with bituminous; 800,000 with subbituminous and lignite	30	45 (at ¼ load)
Front feed—gravity	Traveling or chain	Over 20,000	1½ in. to ¾ in. x 0 nut & slack, not to exceed 35% through ¼ in. round-hole screen	Free burning to medium caking (free swelling index less than 6). Min. ash content (dry) 6%.	500,000 for bituminous; 400,000 for anthracite	20	55 (at ⅛ load)
Front feed—screw or gravity	Vibrating or pulsating	Up to 20,000	1½ in. to ¾ in. x 0 nut & slack, not to exceed 35% through ¼ in. round-hole screen	Free burning to medium caking (free swelling index less than 6). Min. ash content (dry) 6%.	450,000	30	60 (at ⅛ load)

[a] Burning rate is the higher heating value (in Btu) of the type of coal used, multiplied by the number of pounds of coal burned per hour to obtain rated boiler capacity, divided by the total active burning area, in square feet, of the stoker. The maximum values shown in the chart are based on the assumption that furnace walls are water cooled, that there is adequate furnace volume, and that the most desirable type of coal for the unit is used; in absence of these conditions, values should be appropriately reduced to ensure satisfactory combustion.

[b] For complete combustion, there should be no CO in the flue gas, relatively little carbon in the ash discharge (e.g., not more than approximately 15% by weight of total refuse), and only a negligible amount of carbon in the flue gas entering dust collectors or in the stack if dust collectors are not provided.

[c] The values in parentheses are minimum loads (maximum turndowns) for which a stoker should be required to meet the excess-air limitations shown. These values should be met without excessive use of overfire air jets. Where greater turndown than is shown is required for a particular application, it may be specified; however, limitations on excess air percentages and use of overfire air jets should be waived.

[d] Slightly higher ash content is permissible if there is a corresponding increase in ash-softening temperature.

[e] Ash fusion temperature should be stated in contract documents so that the furnace can be designed to give an exit temperature sufficiently low to preclude sooting.

for completing combustion before refuse is discharged to the pit for disposal.

Grates—Grouping of retorts and tuyeres occupies about two-thirds of the combustion surface. The more active burning lanes of the fuel bed are over the tuyere rows of air-admission ports, while the relatively inactive fuel lanes are over the retorts. As the fuel bed completes its movement over the retorts and tuyere banks and reaches the extension-grate section, the burning lanes gradually widen. Constant agitation of the grate section keeps the fuel bed broken up and porous, permitting gradual completion of carbon burning. The ash then moves to the ash-discharge plates where it is ejected for final removal.

Extension grates permit effective burning of coke in the refuse before it reaches the dumping grates. These extension grates are usually equipped with a reciprocating-action drive, either separately or in series with pushers. Agitation of these grates, which may be flat or stepped, keeps the fuel bed active, breaks up any large clinker formation and moves the refuse to the dump grates.

Dump grates, or ash-discharge plates directly aft of the extension grates rid the stoker of refuse after combustion of the fuel is completed. They may be arranged to lower at *intervals* to dump accumulated ash and clinker into the pit, or they may be given a reciprocating motion by connection to the stoker mechanism for *continuous* ash discharge. Air is admitted to most dump grates for a final stage in the combustion cycle before dumping.

Air distribution—Air for combustion is supplied from a plenum chamber (windbox), with the quantity and pressure for the various sections controlled by damp-

ers. Most of the air is admitted through the tuyere blocks. A more elaborate system of controls is usually provided for longer stokers of the gravity-feed type. Pressures for the various sections are indicated by gages, so that closer control may be maintained for greater efficiency. With closer control of combustion air it is possible to obtain higher burning rates without serious increase in maintenance cost.

Stoker drives—The type of mechanical drive for the underfeed stoker depends upon plant load conditions and flexibility desired. Motors are commonly used for low-pressure boilers or in cases where mechanical power is not produced at the plant. Where sufficient steam pressure is available, a steam-turbine or reciprocating-steam-engine drive may be used, particularly when the steam required is a factor in heat-balance conditions. Speed range and flexibility of stoker drives may be increased by variable-speed transmission.

Two types of hydraulic-drive systems are used. In one, a constant-speed motor or turbine drives a variable-displacement oil pump, fitted with proportioning valves, that supplies oil pressure for all drives and controls operating the stoker. In the other, oil pressure from the pump acts directly on the hydraulic cylinder that drives the rams and mechanism for agitating the grates.

Ash discharge—In the single- or double-retort, horizontal-feed stoker, ash discharge is usually at the sides. Ash may accumulate on the side-dump plates to be disposed of periodically or, for units equipped with moving dump-grate sections, it may be discharged continuously into a pit.

In the gravity-feed, multiple-retort stoker, ash is disposed of

at the rear. Dump grates that open intermittently are usually operated by a steam or air cylinder. The period of intermittent dumping will depend upon percentage of ash in the coal and combustion rates per retort. Water sprays are normally installed in the ash pit to cool refuse immediately after dumping. Ash hoppers of ample capacity should be provided for this type of stoker.

Water Cooling

In the early days of stoker application, furnaces were of all-refractory construction. As furnaces became larger and burning rates increased, it became necessary to improve furnace wall and grate construction. This was done by the use of water cooling.

Water-cooled tubes were added to absorb radiant heat and protect the refractory. Water-cooled retorts and tuyeres are also used to lower fuel-bed temperature and chill the ash. This helps to prevent closing of air ports by slag, thereby lengthening life of the parts.

Preheated Air

Development and extensive use of air heaters have effected underfeed-stoker design and performance. Gain in efficiency from preheated air is not confined solely to the heat recovered. With some coals the result has been a thinner and more uniform fuel bed with a decrease in excess air and lower undergrate pressure.

As grate temperature depends upon temperature of combustion air, it is apparent that the latter should not exceed the allowable use limit of the metal of which the grate is made. Generally, the top limit for preheated air, as set by the manufacturer, is 350F.

Table 1 may be used to select stokers when many variables are known.

Ash Handling

Large, modern, space heating and industrial coal burning plants require some clean and economical means for handling ashes. Modern ash handling equipment automatically fulfills these requirements.

The ash content of coals varies to some degree, but for design purposes it is safe to assume that 10% of the coal burned results in ash. It is also assumed that 1 lb of coal produces 10 lb of steam. This efficiency also varies according to the Btu content of the coal and the type of burner used. Table 1 indicates the quantities of ash produced by several industrial steam plants of various capacities operating at full load for one or two 8-hour shifts per day, five days a week.

Ash is usually removed from boiler ash-pits and transferred to a storage silo or bin once every 8-hour shift. The storage silo or bin is designed for a 7-day capacity. To achieve lowest annual owning and operating costs, considerable judgement is necessary in sizing the ash handling equipment. A larger system results in a higher first cost and lower operating cost, whereas a smaller system achieves the opposite result.

There are three types of ash handling systems, *pneumatic, hy-*

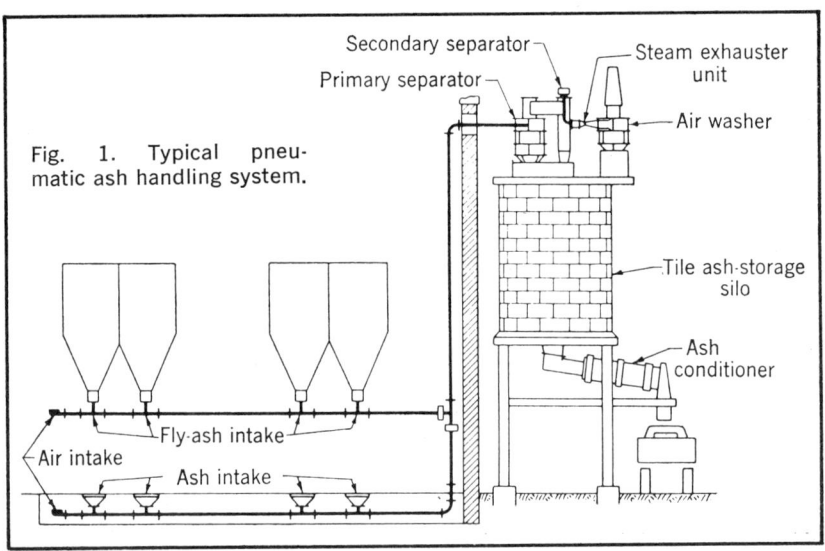

Fig. 1. Typical pneumatic ash handling system.

Table 1. Ash Produced by Steam Plants According to Size

Capacities		Ash Produced		
Full Steam Load, Lb/Hr	Coal Burned, Lb/Hr	Per 8-Hr Shift, Tons	Per 5-Day Week, Tons	
			1 Shift	2 Shifts
200,000	20000	8	40	80
150,000	15000	6	30	60
100,000	10000	4	20	40
50,000	5000	2	10	20
25,000	2500	1	5	10
15,000	1500	0.6	3	6

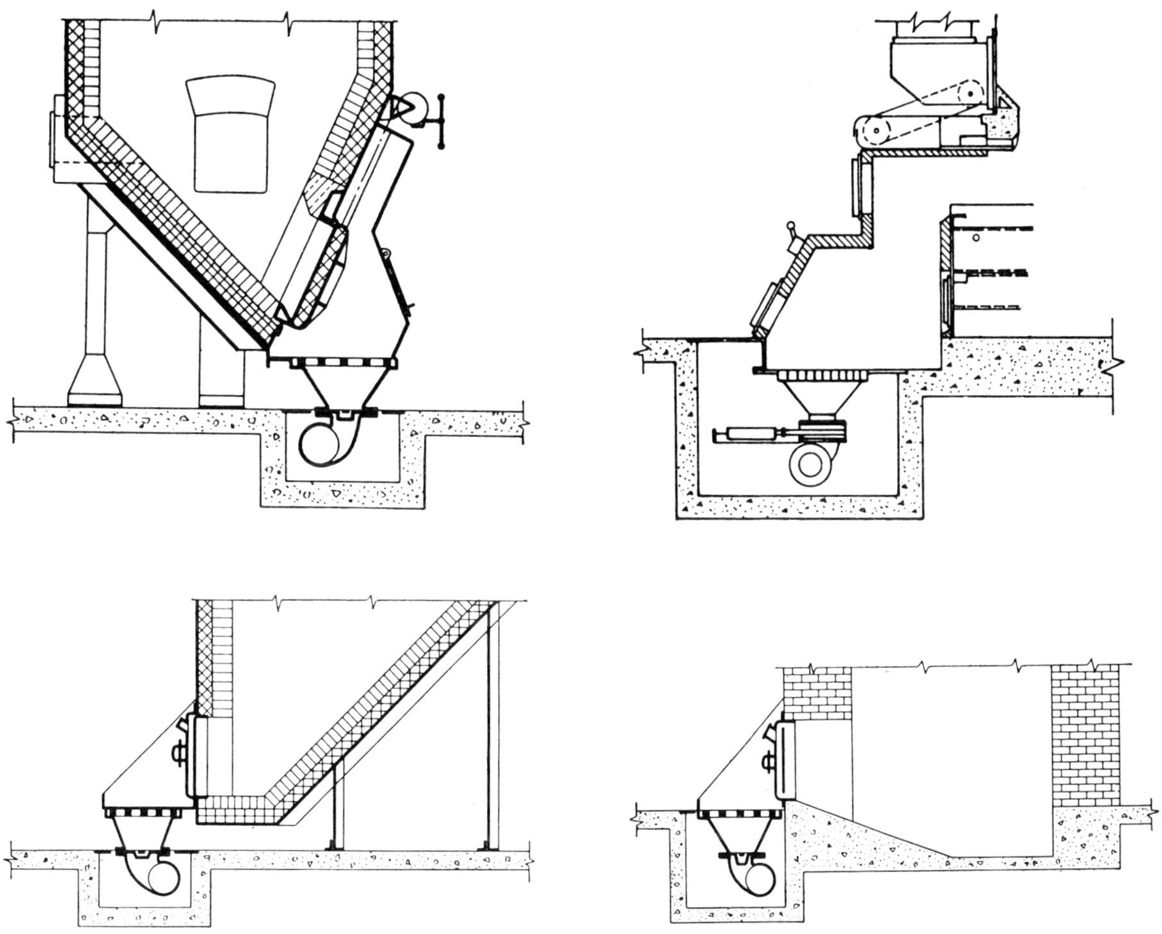

Fig. 2. Four typical designs of bottom ash intakes.

draulic and *mechanical*. As the mechanical ash handling system consists mainly of portable devices and is generally used in plants with capacities below 50,000 lb per hour we will not explore it.

Pneumatic Ash Handling Systems

In pneumatic systems, an air stream produced by a mechanical steam or water exhauster transports ash from boiler ash hoppers and fly ash from dust collectors and traps to silos or bins. These systems are low in first cost and can move ash long distances, horizontally and vertically. They can be used for small or large plants.

Steam or mechanical pneumatic ash-handling systems used in space heating and industrial plants are low in first cost and maintenance. The standard pneumatic system,

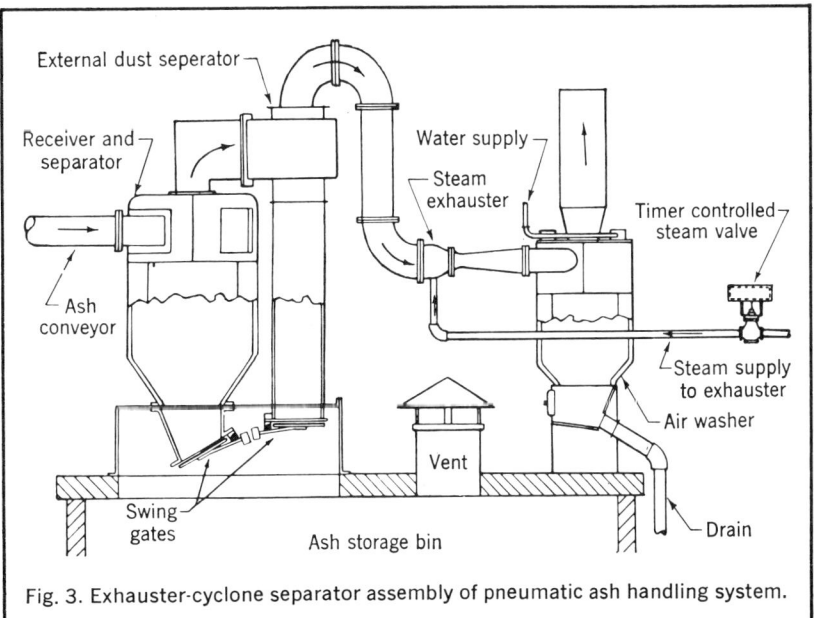

Fig. 3. Exhauster-cyclone separator assembly of pneumatic ash handling system.

Table 2. Standard Silo Sizes

Inside Diameter, Ft	Inside Height	Gross Capacity*		Live Discharge Capacity	
		Cu Ft	Tons	Cu Ft	Tons
8	12' 6"	630	14	460	10
9	14' 5"	920	20	670	15
9	18' 9"	1190	27	890	20
10	19' 9"	1550	35	1130	25
12	18' 9"	2120	48	1340	30
12	20' 10"	2360	53	1580	35
12	22' 11"	2600	59	1820	41
12	25' 0"	2840	64	2060	46
14	22' 11"	3490	79	2230	50
14	27' 1"	4150	94	2890	65
16	27' 1"	5380	121	3560	80
16	31' 3"	6270	141	4450	100
18	33' 7"	8540	190	6700	150
20	36' 4"	11500	260	8900	200
22	38' 5"	14600	330	11100	250
22	43' 7"	16500	370	13300	300

* Capacity in tons is based on ash weighing 45 lb per cu ft.

Fig. 1, is usually installed in plants with full load capacities over 50,000 lb of steam per hour.

Some manufacturers, in an effort to reduce system cost, remove the primary and secondary cyclone separators and substitute a target box. This makes for a very dirty and unsatisfactory system and, although lower in cost, the arrangement should not be employed.

The components of the pneumatic ash handling system are:

Ash Intakes—Ash pits and fly-ash hoppers should have a storage capacity for at least an 8-hour accumulation. Greater storage capacity permits even more flexible ash removal schedules.

Fig. 2 shows typical ash hoppers used with various types of stokers and bottom-ash receivers for ash handling systems. Bottom ash is admitted into the main conveying lines by removing the cast iron plug with a lift hook. At top, right, the intake fitting is equipped with a piston operated slide valve, automatically controlled. Sometimes manual slide valves are used.

Conveyor Piping—Ash conveyor pipe and fittings are made of cast iron, having a Brinnell hardness of 500-550. Pipe should be centrifugally cast. Laterals and elbows should be provided with hand holds. Wear backs on these fittings should have a thickness of 1½-2 in. Hardness should be the same as for pipe. A harder pipe costs more but it lasts much longer. For handling fly-ash, Schedule 80 black steel pipe may be used.

Centrifugally cast pipe comes with plain ends in the following standard wall thicknesses: 4-in. diameter, 0.50 in.; 6-in. diameter, 0.55 in.; 8-in. diameter, 0.65 in.

Exhauster-Cyclone Separator—This unit includes primary and secondary cyclone separators, steam exhauster unit and air washer. Fig. 3 shows a typical assembly. The exhauster induces a vacuum of 5-12 in. of mercury in

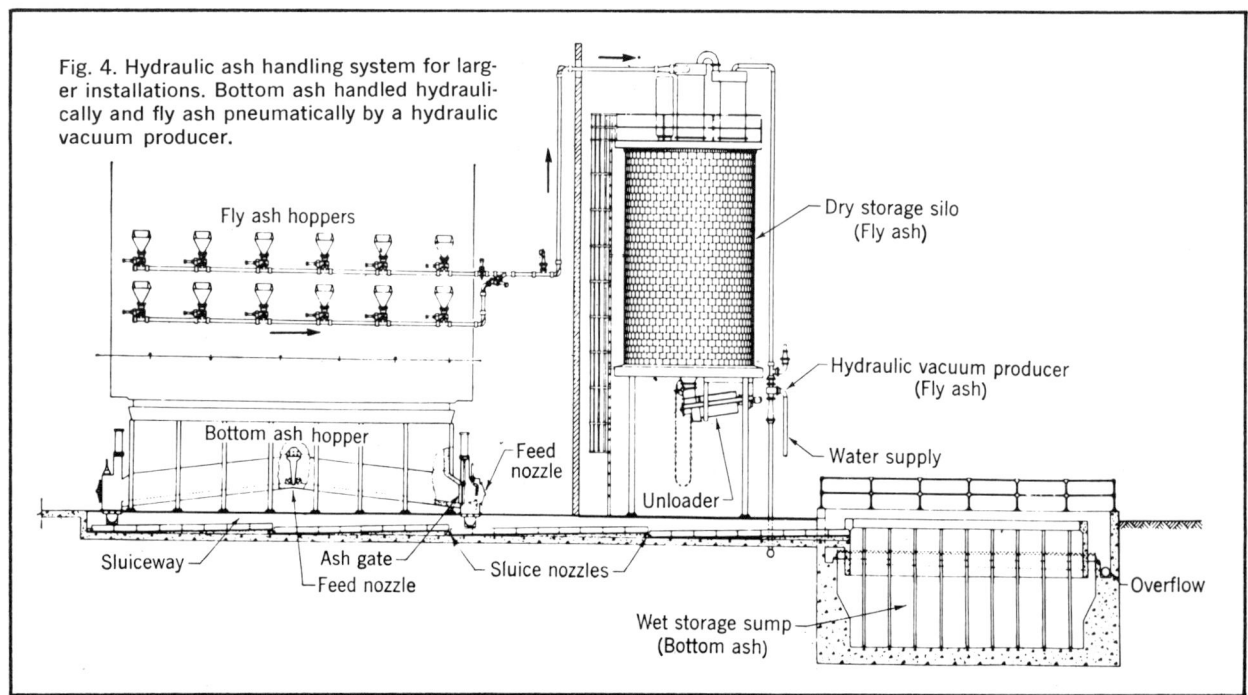

Fig. 4. Hydraulic ash handling system for larger installations. Bottom ash handled hydraulically and fly ash pneumatically by a hydraulic vacuum producer.

Fly ash hoppers

Dry storage silo (Fly ash)

Hydraulic vacuum producer (Fly ash)

Water supply

Bottom ash hopper

Feed nozzle

Unloader

Sluiceway

Ash gate

Feed nozzle

Sluice nozzles

Overflow

Wet storage sump (Bottom ash)

the conveyor lines. The induced primary and secondary separators extract as much as 99% of the entrained ash and dust. An air washer should always be used and the ash dust discharged to the stack. When a mechanical exhauster is used it should be protected by an air washer.

Steam Exhauster—The steam exhauster operates on a minimum steam pressure of 80 psig. Its steam consumption depends upon design of the system, ash characteristic and quantity. As ash systems generally use 6 or 8-in. conveyor lines, the exhauster is line size. Steam consumption for a 6-inch conveyor system can vary from 1500-2500 lb per hour; for an 8-in. system, from 2500-3500 lb per hour. These figures should be used as a guide only. A typical 6-in. pneumatic system with a capacity of 6-14 tons per hour will require approximately 230 lb of steam per ton of ash conveyed.

Mechanical Exhauster—In coal burning plants where high pressure steam is not available to power a steam exhauster, an electric motor driven mechanical exhauster may be used. The ash handling system is the same except that an air washer must be used to protect the exhauster. The cost of the mechanical exhauster is substantially higher than that of the steam exhauster. Exhausters are usually sized to induce velocities of 3000-5000 fpm in the conveyor piping and to create negative pressures that seldom exceed 5 psig below atmospheric.

Ash Storage Silos and Bunkers—Ash storage is generally handled in steel plate or vitrified, glazed tile silos. Tile and steel silos are approximately equal in cost, but the tile silo is favored because of its resistance to corrosion, protection against freezing and low maintenance factor. The silo is the most expensive component of the ash handling system, accounting for 25% to 35% of the total cost. Table 2 indicates standard silo sizes and capacities.

Ash is removed by gravity through segmental discharge gates and a chute to a truck below. Sometimes the chute is furnished with water sprays to reduce dust, but an ash conditioner, though more expensive, is a much better device, as it completely controls all dust. It consists of a motor-driven, rotary drum that mixes ash with water in proper proportions to moisten all dust particles.

Hydraulic Ash-Handling Systems

Hydraulic ash-handling systems use water to transport ash to a dewatering or decantation bin that is periodically emptied. Hydraulic jets sluice ash from the boiler pit into an 8 or 10-in. pipe or trough, and move it along by booster jets to the decantation bin. Fly-ash must be handled separately by steam or mechanical exhauster to the storage silo.

This system may be used in medium to large industrial plants and is much more expensive than the pneumatic types. It is not recommended in areas where water is in short supply or where water pollution is a problem. Details of the system are shown in Fig. 4.

Gas-Oil Burning Equipment, Piping Systems, and Fuels

Fuel oil burning systems consist of burner, combustion air fan, air register, pumping-heating-straining set, piping, accessories, tank heaters and storage tanks.

For brevity, only No. 6 fuel oil will be considered. This fuel varies in viscosity. It must be heated before it can be transferred from the fuel tank to the pump set; then heated again to a higher temperature so that it can be atomized in the burner for clean and efficient burning. A typical fuel flow piping arrangement from tank to burner is shown in Fig. 1 and a tank-type, fuel oil suction heater in Fig. 2.

Types of Fuel Oil Burners

Fuel oil burners are of the gun type. They are set in air registers with manually adjustable vanes for directing a turbulent flow of air to the atomized oil for maximum combustion efficiency.

There are two types of fuel oil burners:

1. A mechanical pressure atomizing type in which fuel oil is supplied to the burner at 200-225F and is atomized for burning at 300 psig. Quantity of fuel burned is controlled by varying oil return line pressure. See Fig. 3.

2. A pressure atomizing, steam-assisted type oil burner that operates at an oil pressure of from 40-150 psig and at 50 psig steam pressure. Fuel oil is supplied to the burner at 190-225F, is atomized when it flows through a sprayer plate and is mixed with steam in the nozzle. This homogeneous mixture then flows through the orifice of the mixing nozzle into the furnace. This type of oil burner is considered the most efficient and economical. See Fig. 4.

Gas Burning Systems

Gas burners are of either the ring or gun type. They are set into the same type of air register as that provided for fuel oil burning systems. See Figs. 5 and 6.

Certain types are designed to be inserted in the air register when only gas is to be burned; others may remain when oil is to be burned. Most burner manufacturers use the gas ring burner unit.

Gas pressures from 8-10 lb at the entrance to the gas train are adequate and provide the correct turn-down ratio for proper burning and control of natural and manufactured gases in large water tube boilers.

Most modified Scotch Marine boilers with capacities up to 25,000 lb of steam per hr are equipped with burners designed to operate with gas pressures from 1-2 lb at the entrance to the gas train.

Many utilities furnish gas at a pressure of from 6-8 in. w.g. A gas booster must then be used to raise the pressure to the proper operating level for burning.

Gas compressors, such as made by Spencer Turbine Co., are driven by electric motors. The cost of operating the compressor must be added to fuel cost.

A gas pressure reducing valve should be used to control the flow of gas to the burner of a steam or HTW generating unit if the gas pressure is greater than the operating pressure of the burners.

Pumping-Heating-Straining Set

Prefabricated and factory built, this unit is constructed on a heavy steel pan. It contains steam and electric fuel oil heaters with automatic temperature controls and fuel oil pumps driven either by electric motors or steam turbines or both, as required. See Fig. 7.

Duplex suction and discharge oil strainers provide clean oil to the burners. Complete with all

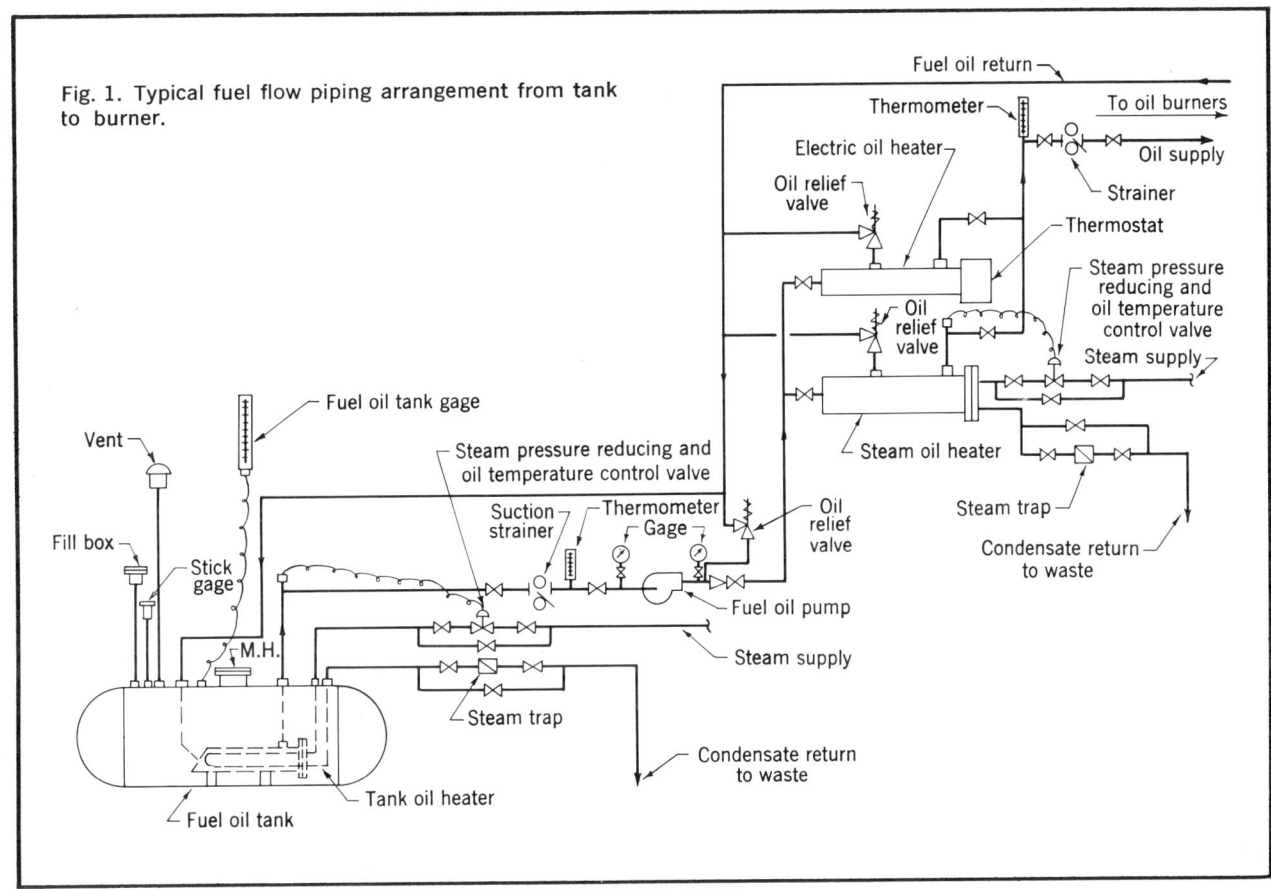

Fig. 1. Typical fuel flow piping arrangement from tank to burner.

necessary manual valves, automatic controls, oil relief valves, pressure gages, thermometers, motor starters and circuit breakers, the unit is furnished piped, wired and ready for installation.

Steam fuel oil heaters are of the shell and tube bayonet type (there are generally two or three). Depending on plant requirements, each is sized to handle full plant load or half load. One heater is always provided for standby.

Steam to these heaters is controlled by self-contained temperature control valves. An electric fuel oil heater can be furnished for cold or emergency startup. It is generally sized to handle one boiler.

Positive displacement type fuel oil pumps are generally of spiral, helical or herringbone gear design. Drives are electric motors or small steam tubines, the latter furnished to provide plant heat balance and breakdown protection.

Usually three pumps are supplied, two for handling the full load and one for reduced loads. They are fitted with relief valves and arranged for manual starting and stopping. These pumps should have the minimum equivalent suction lift of 20 ft. The discharge head should provide 150 psig at a steam assisted burner and 300 psig at a mechanical pressure atomizing burner.

In selecting a pump, a safety factor of 10% should be added to the pump capacity and to the discharge head for wear.

A fuel oil pressure reducing valve is furnished in the discharge header to control oil pressure to the burners.

Suction and discharge strainers are always of the duplex type so that one side can be cleaned while the other is in operation. The set is equipped with powerhouse, mercury type thermometers to indicate oil suction and discharge temperatures. Pressure gages indicate oil suction and discharge pressures as well as steam pressures.

The complete unit is built on a heavy steel oil pan. The pumps are set in the pan with the heaters directly above or racked on one side on structural steel supports. The pan has a valved outlet to draw off any fuel that has leaked from the flanges or pump glands.

The set should be large enough to permit piping to be arranged so that the various components can be replaced or repaired without removing other components.

Fuel Oil Storage

Underground Tanks—Fuel oil is stored underground in horizontally set, welded cylindrical steel tanks in capacities to 30,000 gal. Where total storage exceeds 130,000 gal, it is more economical to provide aboveground vertical cylindrical tanks.

Frequently, subsurface water conditions do not permit the installation of underground tanks. They should never be used where the maximum high water table is above the tank bottom.

Where it is absolutely necessary to bury a tank, it should be anchored to a concrete slab of sufficient weight and the excavation site should be back-filled to overcome flotation.

Steel underground tanks should arrive at the job site with a shop coat of paint to protect them from weather during transit. After the tank is set, it should be wire brushed, and a thick, hot, protective bitumastic coating applied.

Tanks should be Underwriter Approved and of extra heavy construction for long life. Table 1 gives capacities and dimensions of Underwriter Approved, underground oil storage tanks.

Aboveground Tanks—These welded, field erected fuel oil tanks are built of mild, open hearth, high quality, welded steel plates. They should be constructed in accordance with the latest American Petroleum Institute standards for welded fuel oil storage tanks. This construction should also meet National Fire Protection Association (NFPA) requirements.

These tanks are available in capacities from 50,000 to at least 1,050,000 gal. Tanks should be electrically grounded.

Tanks should be set at a level at least 1 ft above the surrounding grade for proper drainage. The tank shell should rest upon a 12-in.-wide, reinforced concrete ring that extends to the proper depth for the load involved. It is assumed here that the bearing load, or the soil, is adequate for the weight of tank and fuel. If not, a substructure must be built.

The area within the concrete ring, from the top down to a distance of 4 in., should consist of well compacted, clean, dry sand, gravel and crushed stone. Before laying the bottom plates, the top layer of dry sand should be covered with sheets of tar paper. Tanks may be protected by steel, concrete or earth dikes. The outer wall and roof should be cleaned and coated with metalic paint.

Local regulations may differ from NFPA requirements and should be checked.

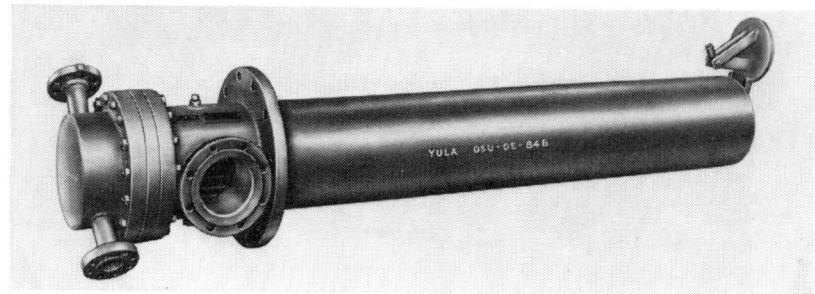

Fig. 2. Tank type, fuel oil suction heater.

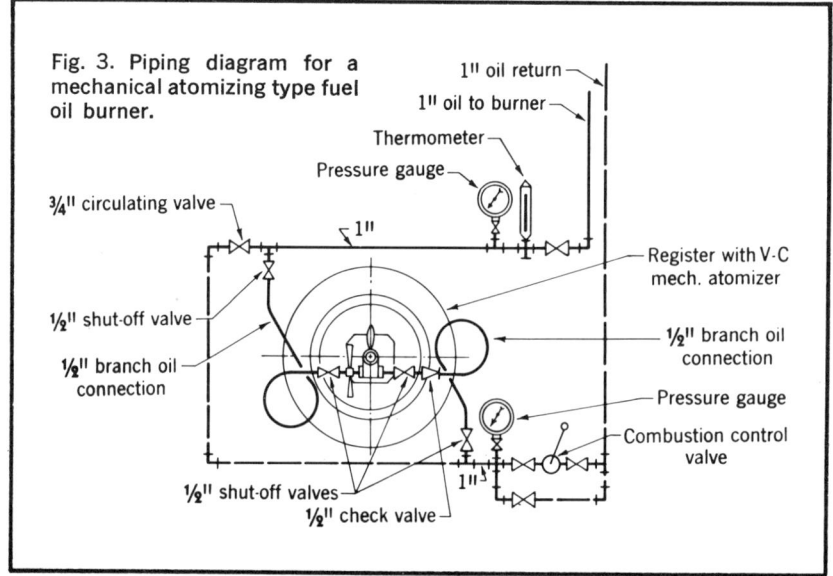

Fig. 3. Piping diagram for a mechanical atomizing type fuel oil burner.

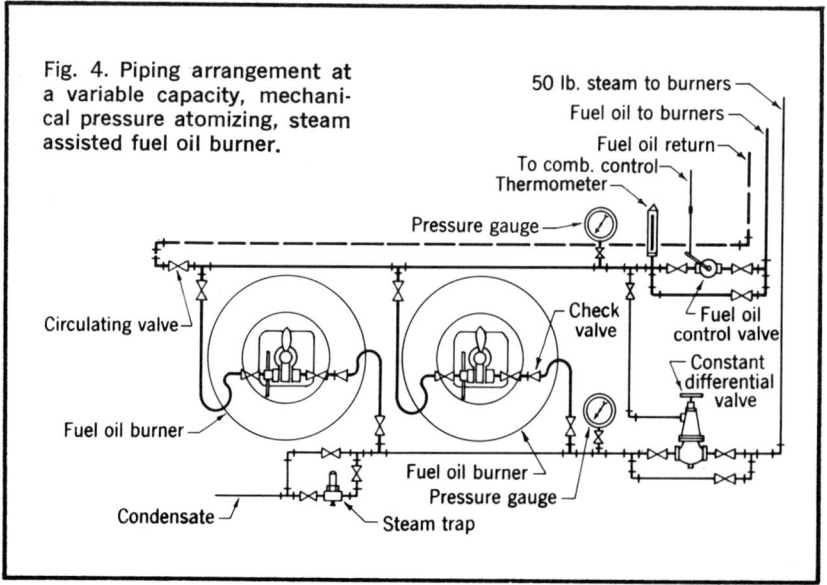

Fig. 4. Piping arrangement at a variable capacity, mechanical pressure atomizing, steam assisted fuel oil burner.

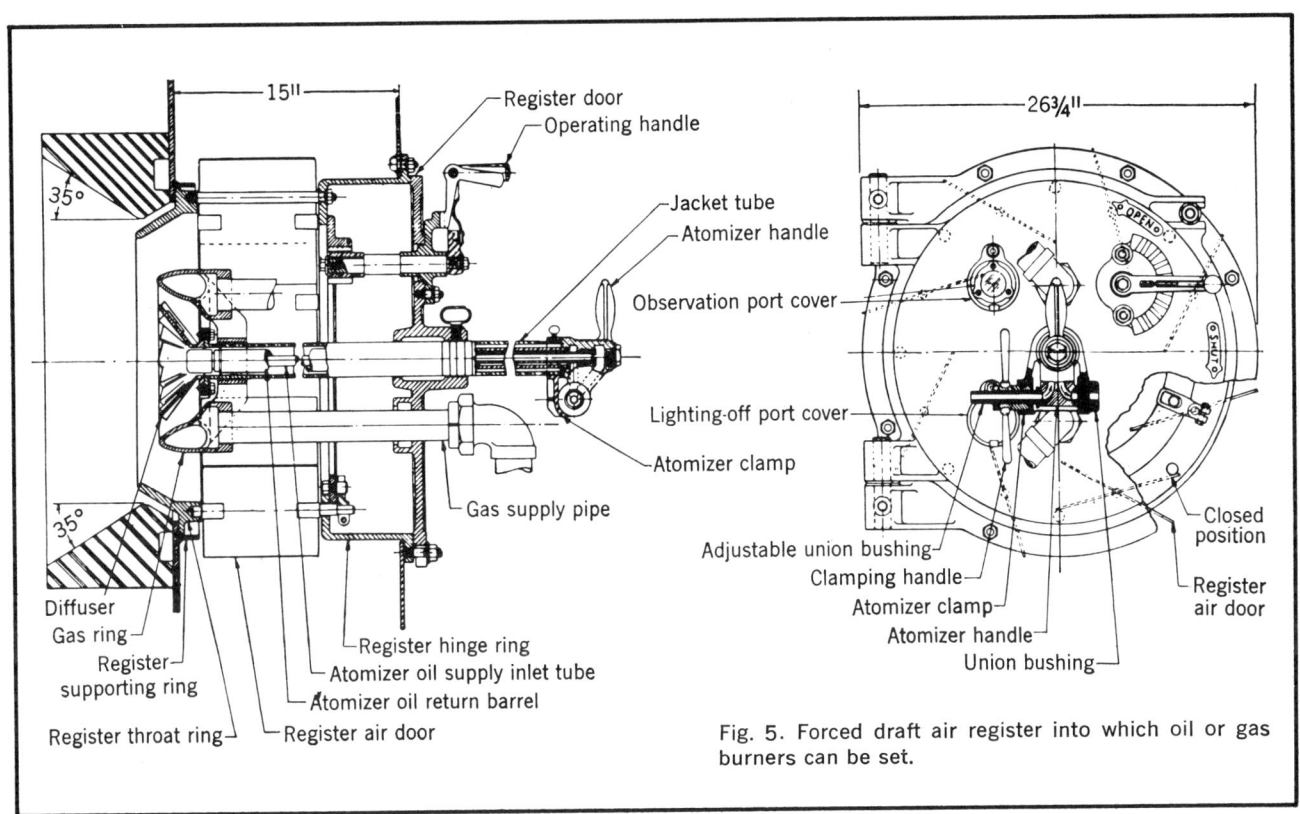

Fig. 5. Forced draft air register into which oil or gas burners can be set.

Table I. Capacities and Dimensions of Underground Oil Storage Tanks (Underwriter Approved, extra heavy, tested to 50 lb pressure)

Tank Capacity, Gal.	Tank Diameter, In.	Length	Shell Size, In.	Dished Heads, In.
550	42	8' 0"	1/4	1/4
550	48	6' 6"	1/4	5/16
1080	48	12' 0"	1/4	5/16
1500	48	16' 8"	1/4	5/16
1500	64	10' 0"	1/4	5/16
2000	64	12' 9"	1/4	5/16
2500	72	12' 9"	1/4	5/16
3000	72	15' 1"	1/4	5/16
4000	72	19' 9"	1/4	5/16
5000	72	24' 6"	1/4	5/16
5000	84	18' 5"	5/16	3/8
5000	96	14' 6"	5/16	3/8
7500	84	27' 0"	5/16	3/8
7500	96	21' 2"	5/16	3/8
10000	84	35' 9"	5/16	3/8
10000	96	27' 7"	5/16	3/8
10000	120	18' 4"	5/16	3/8
12000	96	35' 0"	5/16	3/8
15000	96	41' 11"	5/16	3/8
15000	120	26' 10"	5/16	3/8
20000	120	35' 5"	5/16	3/8
25000	120	43' 11"	3/8	3/8

Concrete Tanks—Concrete fuel oil tanks are often used where the cost of steel plate is prohibitive or where it is not available. They may be constructed above or below ground to almost any capacity.

Tank Heaters

The submerged shell and tube type tank heater has proven the most satisfactory. This unit is of U-tube design and uses steam as the heating medium. It can be furnished with either carbon or stainless steel tubes.

Fuel enters the rear of the shell and passes over the heated tubes before moving into the suction line. The oil return is brought close to the heater opening.

The efficiency of this arrangement lies in the fact that only the quantity of fuel to be pumped is heated, not the whole tank, as would be necessary with grid and helical type coils.

In the underground horizontal type tank, the whole unit is submerged and rests upon structural steel members at the bottom of the tank. Oil suction and return steam and condensate lines are piped up through the top of the tank and properly valved.

In the aboveground vertical tank, the heater is flanged and bolted to the tank wall with the head outside. The whole unit is set just above the bottom of the tank.

Steam, condensate and fuel oil connections are made at the head of the heater and generally run underground to the plant.

Control of the fuel at suction temperature is usually by an automatic, self-contained steam control valve that senses fuel temperature in the suction line after it enters the plant.

When installed outdoors, traps to handle condensate should have a constant bleed to prevent freezing. If installed indoors, an inverted bucket type will suffice. All condensate should, after cooling, be discharged to the sewer to prevent contamination of boiler feed water.

Fuel suction temperature should be maintained at 120F for easy transportation.

Piping

All fuel oil, steam and condensate piping must be insulated. Piping in the plant should be insulated in the usual manner.

Fuel oil, suction, return, steam supply and condensate return between plant and fuel oil storage tanks may be run in steel underground conduit having inner peripheral insulation, concrete box-type trenches or a dry type of *Gilsonite*. The most economical type is a well designed and installed steel conduit. All piping should be welded except the small sizes at the burner and pump set.

Schedule 40 black steel pipe is quite adequate for fuel oil, steam and condensate returns. Condensate piping in conduits and trenches should be Schedule 80.

Fuels

It is not the intent of the author to go into the many and varied details of fuels, as this information is contained in many reference books. From the standpoint of heating plant design, however, proper *selection* of the fuel or fuels is a prime requisite.

The three major fuels used are oil, gas and coal. The volume of usage of other fuels, such as sawdust, bagasse, black liquor, etc., does not compare in any way with oil, gas and coal. In addition, their

Fig. 6. Forced draft air register fitted with gas-electric ignitor, flame safety detection device and burner check housing.

burner and furnace arrangements are of special design.

Average Btu values of the major fuels that may be used for design purposes are: Natural Gas—1000 Btu per cu ft; No. 6 Fuel Oil—

Fig. 7. Fuel oil pumping, heating and straining set.

148,000 Btu per gal or 18,500 Btu per lb; Bituminous Coal—13,000 Btu per lb.

Fuel Selection

Selection of the fuel or fuels to be used depends upon several factors, such as availability, supply, cost, burning ability and storage.

Coal—Bituminous coal is low in cost, and generally high in Btu value. Equipment to handle it, burn it, remove ash and store it is high in first cost and maintenance. Due to air pollution problems, some cities will not allow its use within their boundaries. However, for heating plants outside of cities it is a fine, low-cost fuel, able to meet less rigid air pollution requirements. Highly efficient burners and high draft loss types of fly ash collectors, electrostatic precipitators and wet type scrubbers contribute to cleaner burning and handling of the effluent. The chapters on stokers and pulverizers (ACHV, Dec., 1967) indicate the types of coals that best fit the burning equipment. Complete information on all coals is furnished by the National Coal Association.

Fuel Oil—As No. 6 fuel oil is the only one of commercial importance

Petroleum comes from various parts of the world, high or low in sulphur content depending on the source. During the cracking process, this sulphur ends up in the residual, is released in the steam generator combustion chamber during the burning process and forms a major air pollutant. Air pollution codes of many cities are now limiting the amount of sulphur in fuel oils. The only positive method for removing sulphur and fly-ash is by wet scrubbing flue gases with sodium carbonate in the spray water.

Regardless of the various problems of heating, transportation in the generation of steam, we will not discuss any others here. No. 6 fuel oil is a residual fuel oil, a by-product resulting from the cracking of petroleum to form gasoline, kerosene, naphtha, etc. It is highly viscous and must be heated for transport and burning. Average transport temperature is 120F and average burning temperature, 200F. For proper selection, the viscosity of the oil must be known. There are several standards for measuring viscosity, but the two most frequently used in this country are Saybolt Seconds Universal (SSU) at 100F, and Saybolt Seconds Furol (SSF) at 122F.

and sulphur content, oil is low in price, competing with gas and coal in many areas of the country, and high in Btu content (148,000-154,000 Btu per gal), which accounts for much of its popularity.

Natural Gas—For generating steam in large space heating and industrial plants, natural gas is widely used. It is produced in more than thirty states and distributed at high pressure in pipe lines to most other states. Natural gas is the cleanest of all fired fuels, the easiest to fire and requires no storage or heating. It has an average heating content of about 1000 Btu per cu ft.

Natural gas is sold at two rates. One, the firm rate is higher in cost and the supply is constant; the interruptible rate is lower in cost, but the supply may be interrupted with notice when the supplier's load becomes too large. This generally occurs in cold weather. It means that a plant must have a supplementary fuel to carry it through this period. As such periods are generally of short duration (15 days is an average figure), a No. 4 or No. 2 fuel oil, if available, makes a good standby fuel. The fuel oils are easily stored, with no Btu loss and do not have to be heated.

19

Chimneys and Stacks

The chimney, or stack, performs two functions: it conducts combustion gases to a higher elevation to promote dilution and dispersion, and it produces a draft effect required for combustion. These two functions determine the height of the chimney.

If an induced draft fan is used, or if the steam generator is positively fired, the chimney becomes little more than a vertical duct and a small amount of natural draft is required to lift the gases up the chimney. If, however, the chimney must produce enough natural draft to overcome losses through the passes in a steam generator and breeching, it must be of a carefully calculated diameter and height.

Chimneys, or stacks, may be constructed of radial brick, concrete or steel, with or without refractory lining. The design of brick and concrete chimneys should meet requirements of Associated Factory Fire Insurance Companies.

Radial brick and concrete chimneys are somewhat massive affairs that should be built to withstand hurricane force winds. They must be protected by lightning rods,

grounded and must have a fixed exterior steel ladder for access to the top. They must be furnished with a reinforced concrete foundation. The top should be fitted with a sectional cast-iron cap, to protect the top of the shaft from rain and erosion.

When burning sulphur producing fuels, they must be lined internally for the full height with acidproof brick and set in acidproof mortar.

Concrete chimneys are generally made of reinforced concrete construction.

Steel and Lined Steel Chimneys

Steel chimneys may be self-supporting, free standing or guyed, and lined or unlined with refractory, according to requirements. Guyed chimneys generally used as temporary structures, are a rarity today. If coal or residual fuel is burned in the steam generator, the chimney must be lined with acidproof refractory for its full height. If fuel gas is burned, the lining may be of standard refractory material.

If the chimney rises in a shaft through a building to just above roof level, it must be insulated.

Generally a 2-in. thick, glass fiber blanket, applied over 1-in. deep, "V" rib metal lath is satisfactory for this. The chimney is extended vertically through the building roof and the roof opening arranged with a curb. A steel rain hood is sometimes affixed to the chimney to cover the opening.

Chimney Selection

A chimney may be quickly selected for any plant size with help from the example shown. All data given is for conditions at sea level and 80F outdoor temperature.

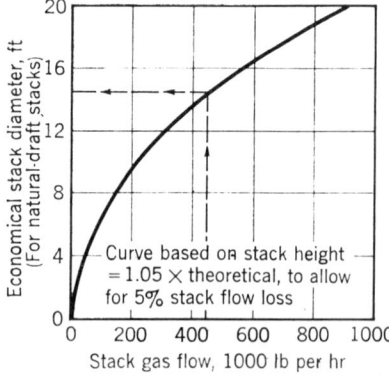

Fig. 1. Economical stack diameter for a range of gas flows.

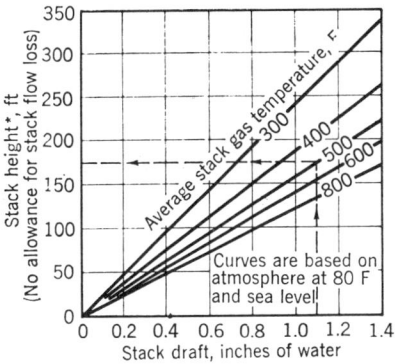

*Stack height is defined as the height above the flue entrance.

Fig. 2. Stack height required for a range of stack drafts and average stack gas temperatures.

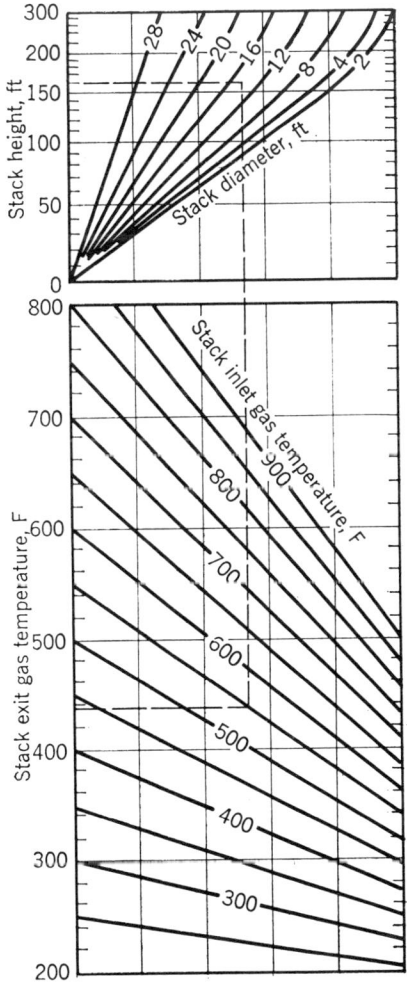

Fig. 3. Approximate relationship between stack exit gas temperature and stack dimension.

Chimney Gas Data

Type of Firing	Approximate Gas Weight, lb per lb of Steam
Oil or gas	1.15
Pulverized coal	1.25
Stoker	1.50

The example shows how to determine economical chimney dimensions.

Given:

Fuel	Pulverized Coal
Steam generated, lb per hr	360,000
Chimney gas flow lb per hr, 360,000 × 1.25 =	450,000
Chimney inlet gas temperature, F	550
Chimney exit gas temperature (assumed), F	450
Chimney draft required (from point of balance to chimney gas entrance), in. of water	1.0
Altitude of plant	sea level

Find:
1. Diameter of chimney to nearest 6-in. increment, ft
2. Active height of chimney, ft
3. Chimney exit gas temperature, F
4. Total height of chimney (add to item 2 as required), ft

Solution:

Chimney Diameter—Enter graph, Fig. 1, with a chimney gas flow of 450,000 lb per hr. This will give a chimney diameter (to the nearest 6-in. increment) of 14 ft, 6 in.

Height (Approximately)—Enter graph, Fig. 2, with the required chimney draft of 1.0 in. of water, and an average gas temperature of (550+450)/2 = 500F. This will give an approximate chimney height of 160 ft.

Exit Gas Temperature—Enter graph, Fig. 3, with the approximate chimney height of 160 ft and diameter of 14 ft, 6 in. Proceed down to the intersection with stack inlet gas temper-

ature of 550F. This will give a chimney exit gas temperature of 436F. The average stack temperature will be (550+436)/2 = 493F.

Height (Actual)—Go back to Fig. 2, with the chimney draft required increased by a 10% factor of safety and, using the average chimney temperature of 493F, determine a new chimney height based on zero chimney flow loss. This will be 177 ft for a chimney flow loss taken at the usual arbitrary value of 5%. The final actual stack height will be 177/0.95 = 185 ft. This is the active height above the flue or breeching entrance, to which must be added any inactive section required from foundation to flue or breeching entrance.

Chimneys Above Sea Level

If the plant is not located at sea level, the chimney draft required should be increased by the altitude factor, approximately 30/B, where 30 is the barometric pressure at sea level, and B is the normal barometric pressure in inches of mercury, at the plant elevation.

Example—Assume that the chimney in the previous example is located 6000 ft above sea level. Normal barometric reading at this elevation is 23.98 in. of mercury from Table 1. Therefore, 30/B = 30/23.98 = 1.25, the altitude factor.

If the chimney height at sea level is 186 ft, the new height will be 186 × 1.25 = 233 ft. This is the height required to produce the same draft effect when the plant is located at an elevation of 6000 ft.

Table 1. Barometric Pressure

Height Above Sea Level, Ft	Pressure,* In. Hg
0	29.92
1,000	28.86
2,000	27.82
3,000	26.81
4,000	25.84
5,000	24.89
6,000	23.98
7,000	23.09
8,000	22.22
9,000	21.38
10,000	20.58

* Values from Bulletin No. 110, 1952, Air Moving and Conditioning Association, Inc.

Scrubbing Flue Gases

For many years, steam and high temperature water plants burning residual fuel oil and bituminous coal have been allowed to discharge via the stack, fly ash, oxides of nitrogen and sulfur dioxide.

But this practice of air pollution is now changing very rapidly, due to federal, state, and municipal air pollution control standards.

These controls and standards require use of low sulfur content fuel oil; bituminous coal is prohibited in some areas; and some method of scrubbing noxious gases and precipitating fly ash is required.

To establish the criteria for a boiler plant scrubber, the designer must ascertain the relevant air pollution codes, type of fuel to be used and the type of equipment needed to obtain compliance with the pollution control standards.

Having established these criteria, the designer should obtain a written guarantee from the scrubber manufacturer that the equipment will perform according to specifications: in terms of percentage of sulfur dioxide,

nitrogen oxides, and amount of fly ash removed, measured in microns. These performance values must be checked by a flue gas analysis when the plant goes into operation.

In general, the two methods of scrubbing gases are (1) dry, and (2) wet.

Dry Scrubbing

Dry scrubbing will suffice where removal of sulfur dioxide and nitrogen oxides is not required and only the fly ash must be captured. This kind of scrubbing can be done effectively by a combination of high draft loss type multiple cyclone and

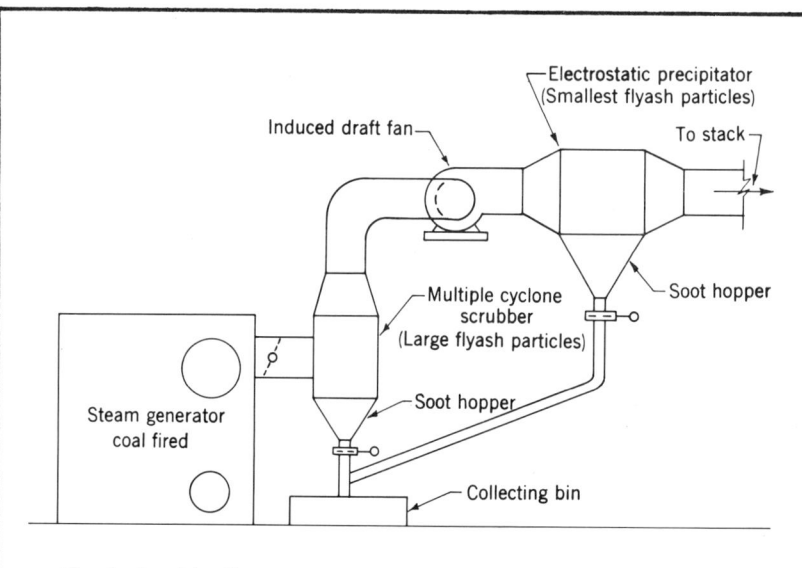

Fig. 1. Combination arrangement of multiple cyclone scrubber and electrostatic precipitator to collect all sizes of fly ash.

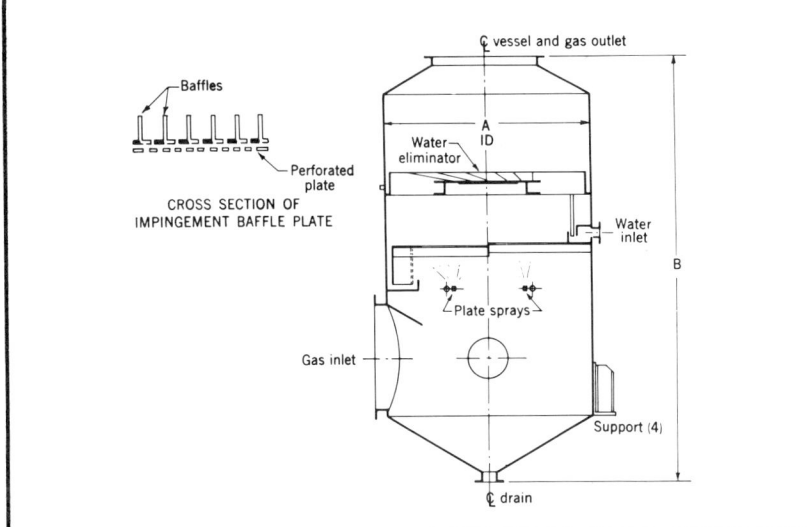

Steaming Rate, lb per hr	Sat. CFM at 20% Excess Air	Inside Diam. A, ft	Overall Height B, ft	Water Flow Rate, gal per min	Na₂CO₃, lb per min
10,000	4,150	3.5	10.2	48	1.1
20,000	8,300	4.5	11.3	65	2.1
30,000	11,700	5.5	13.0	85	3.1
40,000	15,500	6.5	15.4	105	4.2
50,000	19,500	7.0	16.2	118	5.2
60,000	23,500	7.5	16.8	129	6.2
70,000	27,200	8.0	17.6	142	7.3
80,000	31,000	9.0	19.5	170	8.3
90,000	34,900	9.5	20.3	185	9.3
100,000	38,900	10.0	21.0	200	10.3

Notes: 1. All materials in 316 stainless steel.
2. The Na₂CO₃ requirement given in the above chart is the make-up necessary to neutralize the SO₂ in oil fired boilers (S—2.5% by weight).

Fig. 2. Flow diagram and specifications — single stage impingement baffle plate scrubber for oil fired furnace flue gases. Courtesy Peabody Engineering Corp.

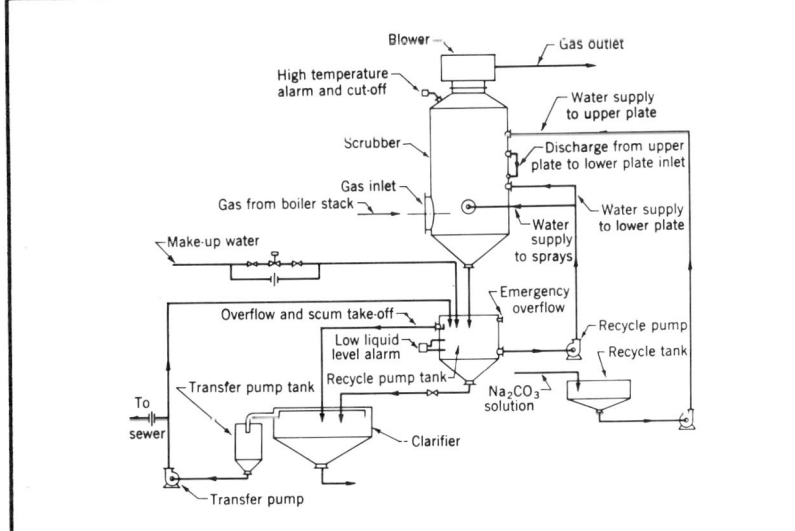

Fig. 3. In this arrangement, clarifier removes sodium sulfite allowing water to be recycled. Courtesy Peabody Engineering Corp.

electrostatic precipitator; the cyclone removes large fly ash particles — the electrostatic precipitator removes the small particles. (See Fig. 1.)

Multiple Cyclone Scrubber

This type of scrubber is usually placed in the breeching at the gas outlet of the steam generator. Multiple cylindrical baffles swirl the gases centrifugally to trap the larger fly ash particles, which collect in hoppers below the cyclones. Hoppers are connected to an ash removal system or, where a water cleaning system is used (to overcome sticky oil ash deposit), the discharge is piped to waste.

To be efficient, draft loss through the gas scrubber must be at least 2.5 inches water gage, a value that requires an induced draft fan to remove combustion gases.

To prevent problems caused by thermal expansion in the breeching, the scrubber should be isolated by expansion joints at the inlet and outlet connections.

If a water cleaning system is supplied, it should be constructed of 316ELC stainless steel.

Electrostatic Precipitator

Fly ash particles too small to be removed by a dry cyclone scrubber can be handled by an electrostatic precipitator. The particles adhere to electrically charged plates and are subsequently removed by mechanical rappers when the unit is shut down.

Although very efficient, it is also the most costly of the dry type scrubbers.

Wet Scrubbing

Wet gas scrubbing is a highly efficient method that removes 90% to 99% of the sulfur dioxide, up to 25% of the nitrogen oxides and reduces particulate matter to 0.03 to 0.04 grains per standard cu ft. Little maintenance is required for such systems - one contributing reason is that construction material is 316ELC stainless.

The three broad categories of wet gas scrubbers are: oil fired steam generator type; coal fired steam generator; and a three-stage unit for oil fired boilers where sea water is readily available. Only the first two categories are described here.

Gas Scrubbers for Oil Fired Steam Generators

A typical design of single-stage wet scrubber for oil fired steam generators is shown in Figs. 2 and 3. The basic components comprise a spray section, single impingement baffle plate and a tangential vane water eliminator. A more sophisticated arrangement uses a clarifier (to remove sodium sulfite solids and particulate), transfer pump, and recycle pump so that spray water can be reused.

Gas flow cycle: heated effluent gases enter bottom of scrubber where they are cooled by a water spray containing a sodium carbonate slurry. The mist captures large dust particles and the sodium carbonate reacts with the sulfur dioxide gas to form sodium sulfite, a solid. Gases then pass through the impingement baffle plate under high velocity (created by an induced draft fan at the scrubber outlet) where more water is used to capture the smaller particles of ash. Then, the cleaned gases flow to a tangential vane water eliminator where centrifugal force removes entrained moisture.

The induced draft fan discharges the scrubbed gases into the stack at an exit temperature of 130 to 140 F.

Water flow cycle: Water for the spray header is supplied as fresh water or from the recycle tank or clarifier. Water to the spray header is pumped under positive pressure; at the impingement baffle plate, water is essentially free flow. After passing across the baffle plate, the

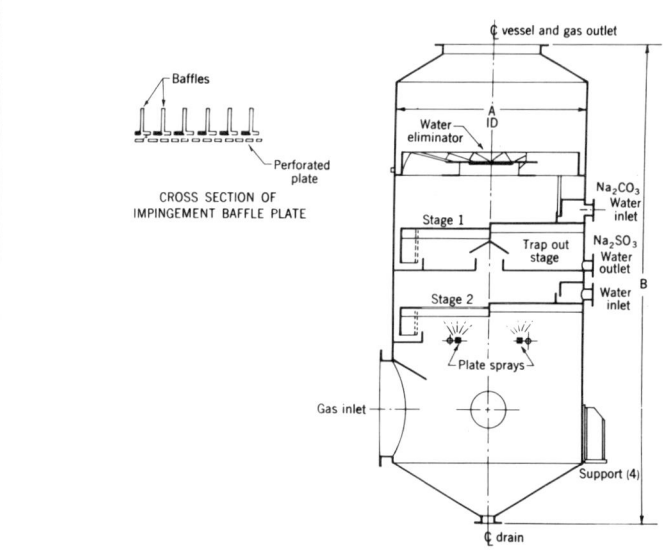

Steaming Rate, lb per hr	Sat. CFM at 30% Excess Air	Inside Diam. A, ft	Overall Height B, ft	Water Flow Rate, gal per min	Na$_2$CO$_3$, lb per min
10,000	4,120	3.5	18.2	48	1.5
20,000	8,260	4.5	19.1	65	2.8
30,000	12,400	5.5	21.0	85	4.3
40,000	16,500	6.5	23.4	105	5.6
50,000	20,600	7.0	24.2	118	7.1
60,000	24,600	8.0	25.7	142	8.4
70,000	28,800	8.5	26.3	155	9.7
80,000	32,800	9.0	27.5	170	11.1
90,000	37,000	9.5	28.3	185	12.5
100,000	41,300	10.0	29.0	200	14.1

Notes: 1. All materials in 316 stainless steel.
2. The Na$_2$CO$_3$ requirement given in the above chart is the make-up necessary to neutralize the SO$_2$ in coal fired boilers (S — 2.5% by weight).

Fig. 4. Flow diagram and specifications — double stage impingement baffle plate scrubber with trap-out stage for coal fired furnace flue gases. Courtesy Peabody Engineering Corp.

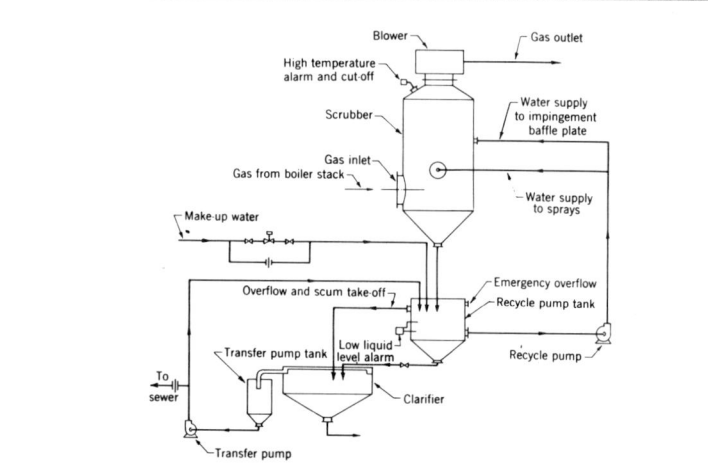

Fig. 5. Addition of clarifier to remove sodium sulfite permits water to be recycled. Courtesy Peabody Engineering Corp.

water drains into the down-comer section and drain where it is pumped into the recycle system where make-up solution is added.

Gas Scrubbers for Coal Fired Steam Generators

A typical split-flow coal fired scrubber design is shown in Figs. 4 and 5.

The main difference between this arrangement and the single-stage design is the use of a two-stage impingement baffle plate design. One pump supplies water and sodium carbonate slurry to the sprays and lower stage of baffles; a second pump supplies water to the second or upper stage of baffles.

Between the two stages of baffles, a trap-out maintains separation between sodium carbonate solution and plain water.

Operating Costs

As an example of operating costs for a wet scrubber, take the figures on the chart of Fig. 4 — the two-stage scrubber for coal fired steam generator. At a steaming rate of 60,000 lb per hour, the corresponding water flow rate will be 142 gpm and the sodium carbonate consumption will be 8.4 lb per min @ 2¢ per lb or $10 per hr. Since water is recycled, the cost of water is only that amount used for make-up.

Of course, to this figure must be added the cost of electric energy for the pumps, fan, etc.

Catwalks, Platforms and Ladders

Access to steam or HTW plant equipment is a necessity. It is accomplished with arrangements of catwalks, platforms and ladders at and around such equipment. These are usually constructed of light steel channels, angles and *I*-beams, with subway grating forming the floors designed to safely support a uniformly distributed live load of 100 lbs per sq. ft. They are supported from the building construction above and, in some cases, partly from the equipment they serve.

Catwalks should never be less than 18 inches wide; a 24-inch wide walk connecting platforms is generally adequate. Platforms should only be large enough to permit servicing of equipment.

All catwalks and platforms should have toe boards at least 4-inches high and railings not less than 42-inches high. Suspension from the building construction above or the equipment itself may be with steel angles or rods. Ship ladders should always be furnished and vertical ladders used only when installation of a ship's ladder is impractical.

Safety chains with hooks should be installed at the head of all ladders or at access points to equipment. Side rails of vertical ladders

should be extended for a distance of 5 ft 10 in. above the platform they serve in order to allow the operator to walk safely on to the platform.

All construction, wherever possible, should be welded. Only those sections that are required to be removable should be bolted.

Typical Specification

a. *General*—All catwalks, platforms and ladders shall be furnished and installed where shown on the drawings.

b. *Design*—The catwalks, platforms and ladders shall be designed to safely support a uniformly distributed live load of 100 lb per sq ft. Design of structural steel shall be based upon a unit working tensile stress of 20,000 psi and shall conform to specifications of AISC and ACI.

c. *Access*—Catwalks, platforms and ladders shall be provided as shown on the drawings or specified to give a safe access to all equipment needing frequent servicing.

d. *Suspension*—Catwalks and platforms shall be constructed with steel stringers suspended

from overhead structural supports and provided with steel floor grating. The stringers shall be formed of steel channels as shown on the drawings. Floor gratings shall be made of 1 in. × ½ in. steel bars assembled on edge with a clear space between bars of not more than ¾ in. The bars shall be bolted to stringers with ⅜-in. machine bolts spaced not more than 2 ft apart.

Walkways shall be suspended from overhead structural supports with steel angles along their sides at the ends of each section and at intermediate intervals of not more than 8 ft. Where overhead supports are impractical, the walkways shall be supported by steel angle struts from the boiler room floor, structural beams or walls of the building.

e. *Toe Boards*—Steel toe boards not less than 4 in. high and at least ³⁄₁₆-in. thick shall be provided on both sides of all walkways and platforms located 8 ft or more above the nearest floor. Toe boards shall be constructed so as to be solid; clips shall be installed for this purpose between the railing posts to prevent any movement.

f. *Ladders* shall be provided where shown on the drawings and shall be ship ladders wherever possible. Vertical ladders shall be used only where ship ladders are impractical. Ship ladders shall be constructed with steel plate stringers and floor grating treads.

Stringers shall be set at an angle of 60° with the floor unless otherwise shown and shall be bolted to the channel stringers. Treads

shall be of the same construction as walkway floor gratings and shall be attached to stringers with 1¼-in. × 1¼-in. × ¼-in. steel angles using ⅜-in. machine bolts.

Vertical ladders shall be constructed using 1½-in. × ⅜-in. steel side bars spaced 18 in. apart and ⅝-in. round steel rungs spaced 12 in. apart. Rungs shall be tapered on ends and securely welded into the side bars.

g. *Railings* shall be provided on both sides of catwalks and platforms except within 3 in. of a wall and on both sides of ship ladders. They shall be constructed with 1¼-in. standard weight steel pipe. Vertical members shall be not more than 8 ft apart. Steel angles for suspension of catwalks may be vertical members of the railing.

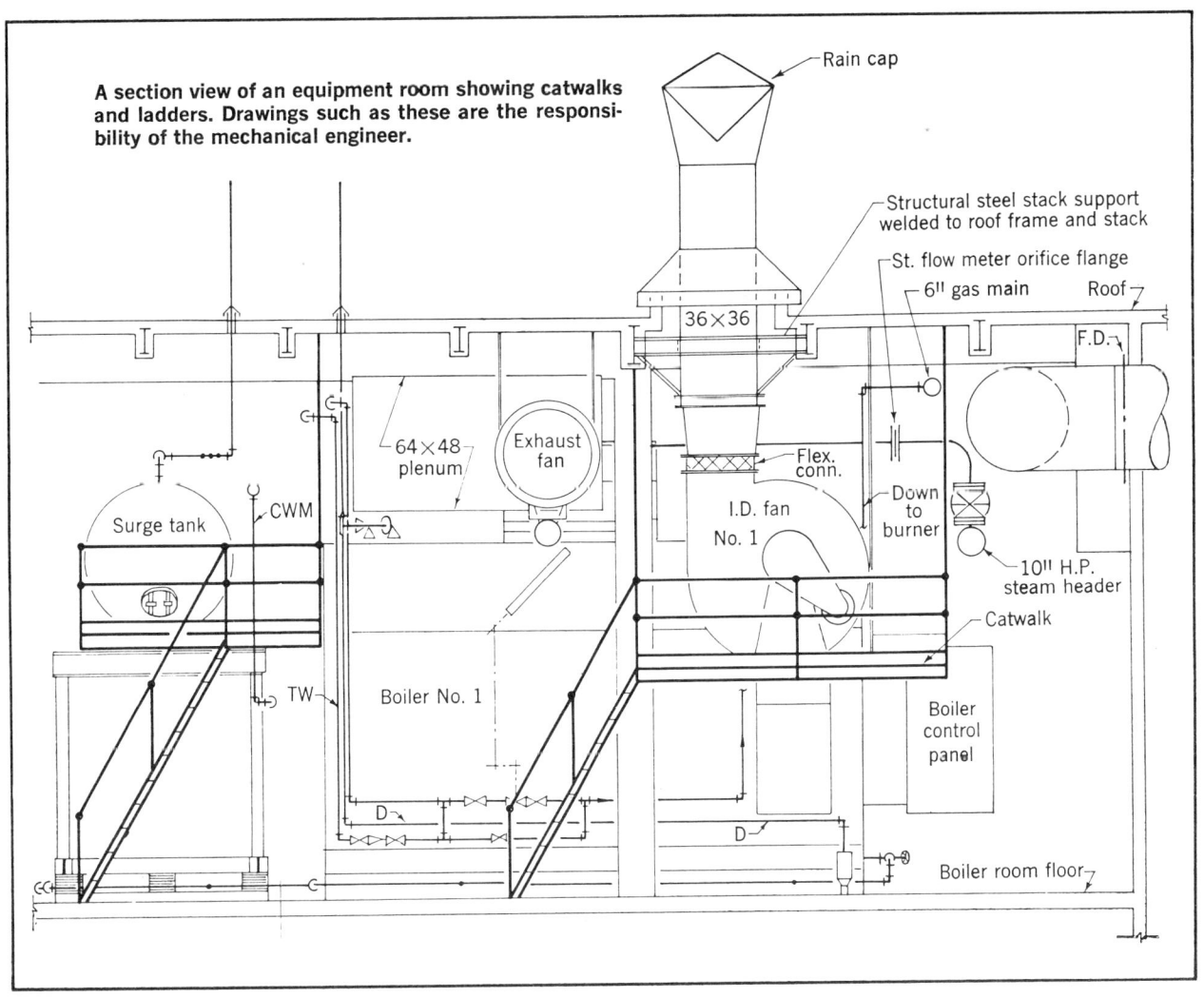

A section view of an equipment room showing catwalks and ladders. Drawings such as these are the responsibility of the mechanical engineer.

The catwalk railing shall consist of two horizontal members; two sloping members shall be used in ship ladders. The upper horizontal member shall be 42 in. above walkway floors or the nosing of treads; the other member shall be located at approximately half of this height. Where openings in railings are provided for portable ladder, two slack chains shall be installed across the openings in place of the rails, one end of each to be attached by a hook.

h. *Connections*—Suspension members shall be attached to overhead structural members of the building. Connections to trusses shall be at panel points only. Any additional structural members required to make the connections herein specified shall be furnished and installed by the contractor. Supplementary steel members shall be attached to building structural members with 2-in. × 2-in. × ¼-in. clip angles or ¼-in. steel plates using ½-in. machine bolts. Suspension angles shall be bolted directly to stringers with ½-in. machine bolts. There shall be two bolts at each connection where possible. Steel ladder members shall be attached to concrete floors with ½-in. machine bolts and expansion shields of the double cinch type.

i. *Shop Painting*—All metal parts of walkways and ladders shall be given a shop coat of red and linseed oil paint.

j. *Finish Painting*—All steel work shall be given a final coat of Rustoleum enamel.

Plant Cleaning, Start-Up, and Tests

During construction, pipe should be kept as clean as possible while under storage, and a real effort should be made to remove dirt, scale and weld metal after installation.

Cleaning

When erection work is completed, piping should be flushed down with cold water (generally this is all that is available) to remove loose dirt, weld metal and scale. If the piping system conveys high temperature water, it should be filled, and sodium carbonate and phosphate added, after which the water temperature should be raised to 200F and maintained ("cooked") at that temperature for at least two days.

While "cooking," the system should be blown-down continuously, (at a slow rate, not greater than the capacity of the make-up pump) at the various blow-down locations to remove dirt and scale. While this is being done, blow-down should be carefully checked, and the reduction of dirt and scale carefully noted.

This procedure may be accomplished simply by blowing-down a water sample into a white enamel pail. When the blow-down is clean, the system should be drained, flushed, and the system water analyzed for chemical content. This flushing should be continued until analysis indicates that the system is chemically clean.

In a steam system, the piping is cleaned by cold flushing, and by circulating steam thru the piping system. Dirt pockets at all drip points are opened, and blown-out until the piping system is clean.

The steam generator is cleaned by using the same chemicals employed for a high temperature water piping system.

During the cleaning operation, the steam generator should be periodically blown-down, and chemicals and make-up water added as required.

Testing

All piping must first be cold tested at a pressure at least one half times its working pressure. This is accomplished by filling the piping with water and carefully venting it to remove all air.

A pressure pump is then connected to the piping and the pressure raised to one and one half times the design working pressure.

The valve between the pressure pump and the piping is then tightly closed, and a reading of the test pressure gage noted.

This test should be run for at least 8 hours, and at the end of the test period, a reading of the pressure gage taken, and compared with the initial test gage reading. If there is no difference in the readings, the cold test should be accepted.

If there is a difference indicating a pressure drop in the pipe line, all welds, valves and flanges must be inspected, and the leak found and repaired. The whole procedure must be repeated until the piping is found to be completely tight.

The final test is with the medium that the piping was designed to contain. The piping should be slowly brought up to its working temperature and pressure. While this is being done, the piping should be carefully checked for leaks, (leaks that might not show under a cold test may become suddenly apparent under escalated temperature and pressure) and piping expansion.

Expansion joints, expansion loops, valves, flanges and welds must be carefully checked. When stability is reached, expansion joints and loops should be checked to see if expansion is within the calculated limits. The attitude of the piping must be scrutinized to make sure that it has not bowed and left its supports in certain areas.

All anchors should be carefully checked to see if they are properly restraining pipe movement. The time for the temperature and pressure test should be at least 8 hours.

Start-Up and Equipment Tests

The HTW system should be brought up to temperature slowly and thoroughly checked for leaks and piping expansion.

Water temperature, pressure, circulation, heat exchangers, safety relief valves, controls, pumps and combustion should also be thoroughly checked for proper operation.

Steam generators and their systems should be started up and checked in the same manner.

The HTW generators and steam generators can then be tested for combustion efficiency on the basis of the ASME short form. These tests should be performed by the mechanical contractor, or by an agency designated by him, and acceptable to the designer.

The tests should be witnessed by the designer.

A 4 hour test of each generator across the specified performance range should be adequate to show that the generators meet the specifications.

Index